Tom Kammermeier

Dilute magnetic semiconductors based on GaN and ZnO

Tom Kammermeier

Dilute magnetic semiconductors based on GaN and ZnO

Structural and magnetic investigation of Gd:GaN and Co:ZnO

Südwestdeutscher Verlag für Hochschulschriften

Impressum/Imprint (nur für Deutschland/ only for Germany)
Bibliografische Information der Deutschen Nationalbibliothek: Die Deutsche Nationalbibliothek verzeichnet diese Publikation in der Deutschen Nationalbibliografie; detaillierte bibliografische Daten sind im Internet über http://dnb.d-nb.de abrufbar.

 Alle in diesem Buch genannten Marken und Produktnamen unterliegen warenzeichen-, marken- oder patentrechtlichem Schutz bzw. sind Warenzeichen oder eingetragene Warenzeichen der jeweiligen Inhaber. Die Wiedergabe von Marken, Produktnamen, Gebrauchsnamen, Handelsnamen, Warenbezeichnungen u.s.w. in diesem Werk berechtigt auch ohne besondere Kennzeichnung nicht zu der Annahme, dass solche Namen im Sinne der Warenzeichen- und Markenschutzgesetzgebung als frei zu betrachten wären und daher von jedermann benutzt werden dürften.

Verlag: Südwestdeutscher Verlag für Hochschulschriften Aktiengesellschaft & Co. KG
Dudweiler Landstr. 99, 66123 Saarbrücken, Deutschland
Telefon +49 681 37 20 271-1, Telefax +49 681 37 20 271-0
Email: info@svh-verlag.de
Zugl.: Universität Duisburg-Essen, Diss., 2010

Herstellung in Deutschland:
Schaltungsdienst Lange o.H.G., Berlin
Books on Demand GmbH, Norderstedt
Reha GmbH, Saarbrücken
Amazon Distribution GmbH, Leipzig
ISBN: 978-3-8381-1717-1

Imprint (only for USA, GB)
Bibliographic information published by the Deutsche Nationalbibliothek: The Deutsche Nationalbibliothek lists this publication in the Deutsche Nationalbibliografie; detailed bibliographic data are available in the Internet at http://dnb.d-nb.de.

 Any brand names and product names mentioned in this book are subject to trademark, brand or patent protection and are trademarks or registered trademarks of their respective holders. The use of brand names, product names, common names, trade names, product descriptions etc. even without a particular marking in this works is in no way to be construed to mean that such names may be regarded as unrestricted in respect of trademark and brand protection legislation and could thus be used by anyone.

Publisher: Südwestdeutscher Verlag für Hochschulschriften Aktiengesellschaft & Co. KG
Dudweiler Landstr. 99, 66123 Saarbrücken, Germany
Phone +49 681 37 20 271-1, Fax +49 681 37 20 271-0
Email: info@svh-verlag.de

Printed in the U.S.A.
Printed in the U.K. by (see last page)
ISBN: 978-3-8381-1717-1

Copyright © 2010 by the author and Südwestdeutscher Verlag für Hochschulschriften Aktiengesellschaft & Co. KG and licensors
All rights reserved. Saarbrücken 2010

to Henrike

Abstract

The two wide band gap dilute magnetic semiconductors (DMS) Gd:GaN and Co:ZnO are among the most favored materials for spintronic applications. Despite intense research efforts during the last years, the origin of the magnetic order is still under debate. This work reports structural and magnetic investigations on these DMS materials employing several complementary techniques. The X-ray linear dichroism (XLD) has been used to gain element-specific insight into the local structure of dopants and cations. X-ray diffraction (XRD) was used to probe the global structural properties. Magnetic characterization by superconducting quantum interference device (SQUID) has been complemented by electron spin resonance (ESR) and X-ray magnetic circular dichroism (XMCD).

Gd:GaN samples were fabricated by focused-ion-beam (FIB) implantation and molecular beam epitaxy (MBE). Room temperature ferromagnetic-like behavior as found for some of our samples by SQUID could not be reliably reproduced. Instead XMCD measurements at the Gd L_3-edge reveal paramagnetic behavior of the dopant. Additionally a possible magnetic polarization of Ga atoms of the host crystal is shown to be too small to explain the total magnetization of these samples. In some samples the formation of Gd and GdN clusters was evidenced by ESR measurements but it can only account for low temperature ferromagnetic-like behavior. Intrinsic room temperature ferromagnetism of this material as seen by SQUID cannot be confirmed by any other technique - neither ESR nor XMCD.

Co:ZnO samples used for this work were predominantly grown by reactive magnetron sputtering (RMS). As shown by XLD analysis, 95% of the Co atoms are incorporated on substitutional Zn-sites in samples of best structural quality. These samples consistently show paramagnetic behavior as found by SQUID, XMCD and ESR. RMS growth of Co:ZnO with reduced oxygen partial pressure yields a magnetic behavior known from ferromagnetic nanoclusters. The X-ray absorption near edge spectroscopy (XANES) and XMCD at the Co K-edge support an increased fraction of Co atoms with metallic character in these samples. A reduced XLD signal indicates less substitutional Co-atoms. These samples were subject to annealing procedures either conducted under O_2 atmosphere or high vacuum (HV) conditions. While the latter strongly enhances ferromagnetic-like properties, they vanish upon O_2 annealing. XANES and XLD analyses show that non-substitutional Co atoms are oxidized to Co_3O_4 by annealing in an O_2 atmosphere, whereas HV annealing increases the fraction of a metallic Co phase. ESR measurements consistently show signatures of superparamagnetic ensembles at elevated temperatures (> 60 K) and isotropic spectra of blocked magnetic moments of nanoparticles at low temperatures. Samples of high structural quality, i.e. with a large fraction of substitutional Co, are inert to annealing procedures.

Zusammenfassung

Die beiden verdünnten magnetischen Halbleiter (DMS), Co:ZnO und Gd:GaN, gehören zu den favorisierten Materialien für mögliche Spintronik-Anwendungen. Trotz intensiver Forschung während der letzten Jahre ist die Art des intrinsischen Magnetismuses dieser Materialen nach wie vor umstritten. Der Gegenstand dieser Arbeit sind strukturelle und magnetische Untersuchungen dieser Materialen mit unterschiedlichen, komplementären Methoden. In Ergänzung der Röntgendiffraktometrie (XRD) wurde der Röntgen-Linear-Dichroismus (XLD) genutzt, um elementspezifisch lokale strukturelle Informationen über Dotieratom und Wirtskation zu erhalten. Die magnetische Charakterisierung mittels SQUID-Magnetometrie wurde durch Messung der Elektronenspinresonanz (ESR) und des Röntgenzirkulardichroismus (XMCD) komplettiert.

Im Falle des **Gd:GaN** wurden Gd-Ionen implantierte wie auch per Molekularstrahlepitaxie (MBE) gewachsene Proben untersucht. Bei 300 K zeigten nur wenige Proben scheinbar ferromagnetisches Verhalten in SQUID-Messungen, welches allerdings nicht zuverlässig reproduziert werden konnte. Stattdessen konnte an der Gd L_3-Kante mittels XMCD paramagnetisches Verhalten des Dotieratoms nachgewiesen werden. Ergänzend konnte gezeigt werden, dass eine mögliche magnetische Polarisation des Ga-Untergitters im Wirtskristall zu gering ist, um die Gesamtmagnetisierung der Probe zu erklären. Somit müssen extrinsische Ursachen für gelegentliche ferromagntische Signaturen in integralen Magnetisierungsmessungen in Betracht gezogen werden. Phasenseparation, verursacht durch Gd und GdN Nanocluster, wurde durch ESR Messungen in einigen Proben nahe gelegt, kann aber ferromagnetisches Verhalten lediglich bei tiefen Temperaturen in dem Material erklären. Intrinsischer Ferromagnetismus bei Raumtemperatur von Gd:GaN konnte weder mit ESR noch mit XMCD bestätigt werden.

Die **Co:ZnO** Proben dieser Arbeit wurden vorwiegend durch reaktives Magnetronsputtern (RMS) hergestellt. 95% der Co Atomen besetzen substitutionelle Zn-Plätze in Proben von höchster struktureller Qualität. Diese Proben zeigen übereinstimmend paramagnetisches Verhalten in SQUID-, XMCD- und ESR-Messungen. RMS Wachstum unter reduziertem O_2-Partialdruck induziert magnetisches Verhalten, wie es von ferromagnetischen Nanoclustern bekannt ist. Röntgennahkantenabsorptionsspektren (XANES) und XMCD bestätigen einen erhöhten metallischen Co Anteil in diesen Proben. Entsprechend ist das XLD Signal, welches ein Maß für substitutionelles Co darstellt, reduziert. Zusätzlich wurden diese Proben unter Sauerstoffatmosphäre oder Hochvakuum (HV) getempert. Während Letzteres die ferromagnetischen Eigenschaften der Proben verstärkt, führt das O_2-Tempern zu einem Verschwinden der ferromagnetischen Signaturen. XANES und XLD Analysen zeigen, dass nichtsubstitutionelle Co Atome bei dem Tempern unter O_2-Atmosphäre zu Co_3O_4 oxidiert werden, wohingegen HV-Tempern eine Erhöhung des metallischen Co Anteils bewirkt. ESR Messungen bei erhöhter Temperatur (> 60 K) zeigen übereinstimmend Spektren, welche für superparamagnetische Ensembles charakteristisch sind. Isotrope ESR-Spektren von geblockten magnetischen Momenten von Nanopartikeln werden hingegen bei tiefen Temperaturen beobachtet. Proben von hoher struktureller Qualität, d.h. einem hohen Anteil substitutioneller Co Atome, sind inert gegenüber Temperprozeduren.

Contents

Abstract/Zusammenfassung

Introduction   1

1 Theoretical background   5
  1.1 Basics of magnetism 5
  1.2 Wide band gap semiconductors ZnO and GaN 8
      1.2.1 Wurtzite crystal structure of ZnO 9
      1.2.2 Wurtzite crystal structure of GaN 11
  1.3 Electron magnetic resonance 11
      1.3.1 The splitting of Co^{2+}-ion states in ZnO 15
      1.3.2 Powderspectra 17
      1.3.3 Ferromagnetic resonance (FMR) 19
      1.3.4 Transition from ESR to FMR 21
  1.4 Element specific investigation methods 23
      1.4.1 X-ray absorption spectroscopy (XAS) 24
      1.4.2 X-ray linear dichroism (XLD) 26
      1.4.3 X-ray magnetic circular dichroism (XMCD) 27

2 Experimental techniques   29
  2.1 Growth of ZnO and GaN based DMS 29
      2.1.1 Reactive magnetron sputtering 30
      2.1.2 Additional growth techniques 30
  2.2 Setup for electron magnetic resonance measurements 32
      2.2.1 Evaluation of ESR spectra 35
  2.3 Superconducting quantum interference device (SQUID) 39
  2.4 XAS measurements at beamline ID12 45

3 Paramagnetic impurities in SiC, ZnO and Al$_2$O$_3$ — 51
3.1 Nitrogen in SiC — 53
3.2 Transition metall ions in ZnO — 54
3.3 Cr^{3+} and Mo^{3+} in sapphire - Al$_2$O$_3$ — 60
3.4 Discussion: Impurities — 63

4 Experimental results for Gd:GaN — 65
4.1 Preparation of Gd:GaN samples — 65
4.2 Paramagnetic signatures in Gd:GaN — 66
4.2.1 Gd:GaN grown on SiC — 66
4.2.2 Gd:GaN grown on sapphire (0001) — 71
4.2.3 Element specific investigations of Gd:GaN — 75
4.3 Matrix polarization and magnetic polarons — 78
4.4 Phase separation and clustering — 81
4.4.1 Indications of phase separation in Gd:GaN — 81
4.4.2 Gd:GaN with evidenced clusters of GdN — 87
4.5 Discussion: Gd:GaN — 95

5 Experimental results for Co:ZnO — 99
5.1 Preparation of Co:ZnO samples — 99
5.2 Co:ZnO - samples of high structural quality — 100
5.2.1 ZnO doped with 10% Co — 100
5.2.2 Co$^+$-ion implantation in ZnO - low dopant concentration — 111
5.3 Co/CoO nanoparticles — 112
5.4 Clustering in Co:ZnO — 116
5.4.1 Clustering in Co:ZnO grown under oxygen deficiency — 116
5.4.2 Metallic Co precipitations in Co:ZnO — 124
5.5 Annealing effects — 131
5.5.1 High vacuum annealing of best quality samples — 132
5.5.2 Annealing of samples with reduced oxygen content — 136
5.6 (Blocked) Superparamagnetic powder interpretation of isotropic ESR spectra contributions — 151
5.7 Discussion: Co:ZnO — 156

Summary — 158

Outlook	162
Appendix	165
A Calculation of the Co^{2+}-ion in ZnO	165
B Sample overview	171
C Preparation Chamber	175
D Heater system	177
E Bake out system	181
Bibliography	186
List of publications	201
List of acronyms	205
Acknowledgement	206

Introduction

Considering the impact of technological progress on every day's life during the last decades, the micro-technological revolution – in particular the miniaturization of transistors – is undoubtedly the most impressive development, since it has led to the computerization of our environment. It is associated with enormous advances in the understanding and fabrication of semiconductor structures. Up to now the electron's charge is the only physical property utilized for data processing. The use of the electron's second fundamental property – the spin – has been limited to data storage where ferromagnetic materials have been used since the twenties of the last century. Also these applications have undergone an impressive development concerning data density from data tapes over floppy discs to hard disc drives during the last decades. The capacity of the latter has increased by almost two orders of magnitude every ten years since 1980 [1, 2].

With the amazing increase of bit density on hard disc drives the electron spin has come more into the focus. One can consider the giant magnetoresistance (GMR)-based read heads as a first commercial success of a technological field nowadays known as spintronics. Consequently A. Fert and P. Grünberg have received the Nobel prize for the discovery of the GMR [3, 4] effect in 2007. A related effect – the tunneling magnetoresistance (TMR) – has resulted in a further improvement of the performance of modern hard discs [5, 6].

The enhanced miniaturization of micro electronics has lead to very high energy densities on today's integrated circuits making cooling a severe issue. Extrapolation of Moore's law [7], which actually describes the miniaturization phenomenologically, reaches inevitably fundamental physical barriers [8]. The prospect to overcome some of these by low power spin-based data processing has driven numerous research efforts during the last years.

Several suggestions to implement the electron's spin in electronics have been reported since the discovery of the GMR [9, 10]. Besides the reduced power consumption of spin-based electronics also unprecedented flexibility of logic devices is envisionaged [11].

One of the major challenges of implementing the spin into processing circuits is the transfer of spin polarization from one material to another. Successful spin injection between metals has been demonstrated by Johnson and Silsbee et al. [12, 13]. The first injection of spin polarized electrons into a semiconductor at room temperature was reported by Zhou et al. [14]. Effects of spin polarization on radiative carrier recombination in a light emitting diode (LED) was demonstrated by Hanbicki et al. [15].

However, one of the remaining problems of a potential spintronic device is the conservation of a spin polarization within a semiconductor. Naturally, a ferromagnetic semiconductor with intrinsic spin polarization would be the material of choice.

Such ferromagnetic semiconductors are known since the sixties of the last century. The first report by Matthias et al. [16] demonstrated ferromagnetic behavior of EuO below a Curie-temperature (T_C) of 70 K, certainly not suitable for device application. Nowadays it is known as one of the rare semiconducting oxides exhibiting ferromagnetism. It is considered to be a narrow bandgap (1.2 eV) semiconductor, which crystallizes in the rocksalt (fcc) structure with a lattice constant of 5.144 Å [17].

Higher T_C's are reported for dilute magnetic semiconductors (DMS) where the dopant induces a carrier mediated long range magnetic order [18]. Mn doped GaAs can be considered to be one of the best studied materials [19] with a T_C of up to 173 K [20].

Within the search for room temperature ferromagnetism in magnetically doped semiconductors a work of Dietl et al. [21] has sparked considerable research efforts on semiconductors with band gaps larger than 3 eV, especially GaN and ZnO. Wide band gap semiconductors have already attracted a lot of interest apart form potential use as DMS. Sophisticated doping of – in particular GaN – has recently led to light emitting diodes of unprecedented performance. The large energy separation between valence band and conduction band results in very low dark currents and thermal leakage in respective devices. Electron and hole mobilities are significantly altered compared to narrow band gap semiconductors. High power amplifiers, CCD devices, lasers and diodes can benefit from these properties.

Two types of wide band gap semiconductors are at the focus of this work, namely

INTRODUCTION 3

the III-V semiconductor GaN and the II-VI semiconductor ZnO. In case of GaN the rare earth ferromagnet Gd is chosen as dopant whereas for ZnO the 3d ferromagnet Co is used as dopant. Both materials Gd:GaN and Co:ZnO [1] have been frequently reported to exhibit the desired ferromagnetic properties at 300 K. First claims date back to the year 2002 and 2001, respectively [22, 23]. Despite these reports of ferromagnetism at room temperature in both types of DMS, eight years later the claims still remain highly controversial [24, 25] and applications in first devices are still lacking.

All Gd:GaN samples used in this work were provided by collaborators. Growth details are given in the appendix or, if necessary, in the respective chapter. Co:ZnO films were predominantly grown by reactive magnetron sputtering. Either by sputtering from composite targets or by simultaneous sputtering from a triple magnetron cluster. Additionally, Co:ZnO samples grown by different methods were provided by collaborators. Details are given in the appendix B.

Most of the early works on Gd:GaN and Co:ZnO concentrate on the growth and structural characterization of the new materials. The magnetic investigation is typically limited to superconducting quantum interference device (SQUID) measurements [26, 27], which at that time were considered to be sufficient to undoubtedly prove ferromagnetism.

The goal of this work is to clarify the origin and possible intrinsic mechanism of the magnetism of theses materials. To account for this, both materials have been investigated much more comprehensively. The SQUID magnetometry is complemented by magnetic resonance and X-ray magnetic circular dichroism (XMCD) measurements. The standard structural investigations by X-ray diffraction (XRD) have been extended by X-ray absorption near edge spectroscopy (XANES) measurements conducted with linear polarized light, exploiting the X-ray linear dichroism (XLD).

Besides the intrinsic element specificity of synchrotron based spectroscopies the latter method offers insight into the local environment of the probed atoms [28] – an information which can hardly be achieved by other techniques.

Synchrotron light was also used for magnetic characterization by XMCD measurements. Like the XLD investigations the XMCD was predominantly measured at the K-edge of respective elements. The disadvantage of probing a smaller signal

---

[1] Throughout this work the notation "X:Y" is used for a material Y doped with X.

than at the L-edge was compensated by the bulk sensitivity of hard X-rays. The latter enables information gain from the entire sample – especially in conjunction with integral SQUID data – in contrast to surface sensitive soft X-ray measurements. Magnetic resonance was used as a second integral magnetization measurement method for this work. Intense scientific efforts have been undertaken to understand the spin's behavior under various conditions, since the experimental verification of it by the Stern-Gerlach experiment [29]. The magnetic resonance measurements have been one of the most powerful tools of magnetic examination for a long time. Numerous measurements have been conducted on transition metal doped host crystals in the mid of the last century. Thanks to sophisticated simulation tools it is nowadays even used in Biology to understand complex molecules. By use of a cavity the sensitivity of this method can be enhanced by orders of magnitude. Even a single atomic layer of a ferromagnetic material [30] can be detected. Recently developed local methods can sense down to 10^6 spins [31]. The high sensitivity makes magnetic resonance a valuable tool for the verification of ferromagnetism. Magnetic resonance measurements on DMS are reported for example for Mn:GaAs [32].

This work is structured as follows: In the first chapter some general remarks and theoretical background will be given. Naturally the treatise will be fragmentary and limited to mentioning the most important points. A second chapter introduces the experimental techniques used in this work in more detail. Some specialized knowledge which will be important in the discussion sections is provided. The third chapter will present experimental results on non-interacting paramagnetic impurities in substrates or host crystals which can be interpreted as limit of ultimate dilution in DMS. The results of both materials, Gd:GaN and Co:ZnO, are treated in individual chapters in this work, respectively. Finally, a summary and a short outlook is given compiling the most important results and presenting some new questions arising form the conclusions of this work.

Chapter 1

Theoretical background

The purpose of this chapter is to provide the reader with the most essential background knowledge needed for this work. A comprehensive introduction into these fields of science is beyond the scope of this work. The reader is referred to the given references for further studies.

1.1 Basics of magnetism

It is still a wide spread misconception that the rules of quantum mechanics do not influence our daily life. Magnetism might be the best example to prove the opposite, considering the Bohr-van Leeuwen theorem which states the non-existence of it if only classical equations are considered [33].
The three basic types of magnetism are best distinguished by the response of a material to an external applied magnetic field, which is described by the susceptibility χ.

$$\chi = \frac{\partial M}{\partial H} \qquad (1.1)$$

Inside a material inner fields add to the applied field H; therefore one defines the magnetic induction B:

$$B = \mu_0 (M + H) = \mu_0 (\chi H + H) = \mu_0 \underbrace{(1 + \chi)}_{\mu} H \qquad (1.2)$$

With μ_0 being the vacuum permeability and μ the permeability of the material. The coexistence of magnetic fields H and B are a constant source of misconceptions and need clarification even nowadays [34].
For **diamagnetism** χ is negative which can be phenomenologically understood as

the reduction of the outer field by magnetic moments caused by induced currents. Even though this perception is actually wrong [35] one can derive a formula for the diamagnetic moment of an atom identical to the quantum mechanical solution. The Langevin diamagnetic moment is given by [36]:

$$\mu_{dia} = -\frac{e^2 <r^2>}{6m_e c^2} B \tag{1.3}$$

With r being the radial coordinate of the electron, e the electron charge, m_e the electron mass, c speed of light and B the applied field.

Diamagnetism is intrinsic to all atoms. A perfect diamagnetic response is realized by superconductors, where $\chi = -1$, which describes the complete expelling of the external field by building up an opposite field [1]. If diamagnetism is not apparent in magnetization measurements the effect is covered by other types of magnetism.

Paramagnetism describes the alignment of existing magnetic moments (e. g. from unpaired electrons) in the direction of the external field. The resulting B-field is enhanced and therefore $\chi > 0$.

From the Hamiltonian of an atom in a magnetic field one can derive the paramagnetic contribution:

$$H_{para} = \mu_B g_J \vec{J}\vec{B} \tag{1.4}$$

g_J is typically calculated by the Landé formula, which assumes **S**, **L**, and **J** to commute with H (Russel-Saunders coupling).

$$g_J = 1 + \frac{J(J+1) + S(S+1) - L(L+1)}{2J(J+1)} \tag{1.5}$$

The quantum number m_J of the angular momentum **J** can vary between -J and J. This leads to a description of the system by a partition function as it is known from statistical physics:

$$Z = \sum_{m_J=-J}^{J} \exp\left(-m_J \frac{g_J \mu_B B}{k_B T}\right) \tag{1.6}$$

The free energy of the system is thus:

$$F = E - TS = Nk_B T \ln Z \tag{1.7}$$

---
[1]This strictly only accounts for so called type I superconductors.

BASICS OF MAGNETISM

With E being energy, T temperature, S entropy and N the number of particles. This can be used to calculate M.

$$M = \frac{-1}{V}\frac{\partial F}{\partial B} = \frac{-1}{V}\frac{\partial F}{\partial Z}\frac{\partial Z}{\partial y}\frac{\partial y}{\partial B} = \ldots = M_S B(y) \qquad (1.8)$$

Where M_S is the saturation magnetization of the system and $B(y)$ the Brillouin function:

$$B(J) = \frac{2J+1}{2J}\coth\left(\frac{2J+1}{2J}y\right) - \frac{1}{2J}\coth\left(\frac{y}{2J}\right) \qquad (1.9)$$

The argument y contains the total angular momentum J and is defined as:

$$y = \frac{g_J \mu_B J B}{k_B T} \qquad (1.10)$$

Note that for $T = 0$ equation 1.9 results in an immediate saturation of the magnetization of a paramagnet in case of an external magnetic field.

The detailed derivation of the Brillouin function makes use of a conversion of the partition sum into a geometric series and can be found in standard textbooks [37]. In the limit of $J \to \infty$ equation 1.9 converges towards the Langevin function:

$$L(y) = \coth(y) - \frac{1}{y} \qquad (1.11)$$

The magnetic behavior of single domain nanoparticles is typically described using the Langevin function in conjunction with **superparamagnetism**.

For the following chapters it is important to stress that the Brillouin function fully explains the magnetization behavior and its temperature dependence of a perfect paramagnetic atomic system. The temperature dependence in the limit of $y \to 0$ ($g_J \mu_B J B \ll k_B T$) is known as Curie law:

$$\chi(T) \propto \frac{1}{T} \qquad (1.12)$$

The term "perfect paramagnetic atomic system" implies a large separation of the ground state of the atom to the excited electron states. If this is not the case excited electron states with different J can perturb the ground state leading to a paramagnetic behavior. The respective paramagnetism was first described by Van Vleck [36]:

$$\chi_{VanVleck} \propto \sum_{m=1,2..} \frac{<m|L_z + gS_z|0>}{E_m - E_0} > 0 \qquad (1.13)$$

Besides types of dia- and paramagnetism mentioned so far there are special variants of magnetism of the conduction electrons in metallic systems (i. e. Pauli Paramagnetism) which will be not described here since they are supposed to play a minor role for DMS. Typically $|\chi| \ll 1$ for dia- as well as paramagnetism.

The latter changes for **ferromagnetism** (and ferrimagnetism). The competition between Coulomb force and the Pauli exclusion principle can result in long range order of magnetic moments within the material. This mechanism can lead to χ in the order of 10^3 to 10^4 and therefore completely dominates the other types of magnetism at elevated temperatures. Respective M(H) curves present no longer a functional dependency, since they depend on the *magnetic history* of the sample. This hysteretic behavior changes above a characteristic temperature – the Curie temperature (T_C) – where thermal excitations result in a breakdown of the long range magnetic order. In the special case of **antiferromagnetism** the exchange interaction leads to neutralization of existing magnetic moments by antiparallel alignment which results in a zero macroscopic magnetization.

The temperature dependent magnetism governed by exchange coupling is characterized by the order temperature. At higher temperatures thermal excitations become stronger than the coupling and the materials become paramagnetic [2].

1.2 Wide band gap semiconductors ZnO and GaN

The two compound semiconductors GaN and ZnO have already been subject of intense research efforts because of their advantageous properties for optoelectronic devices. In particular the progress of LED should be mentioned. Besides the wide band gap of both semiconductors of $\approx$ 3.4 eV [38, 39] and a similar crystal structure, some differences beyond the different valences of the constituents should be mentioned.

In consideration of proposed mechanisms ("carrier mediated") of the magnetism of DMS like Gd:GaN and Co:ZnO the almost three times higher exciton binding energy of ZnO (60 meV) compared to GaN (24 meV) has to be mentioned [40, 41]. In addition the high quality bulk substrates and ease of wet etching of ZnO make it preferable to GaN [42]. However, the most relevant obstacle for spintronic applications of ZnO is the problem of the reliable p-type doping, even though it has been already claimed in 2004 [43]. Up to now the reason for the n-type character of ZnO is under debate. Hydrogen atoms introducing shallow donor states have been recently reported to severely hamper p-type doping [44]. The fabrication of GaN did

---

[2]Curie-temperature (ferromagnet), Néel-temperature (antiferromagnet)

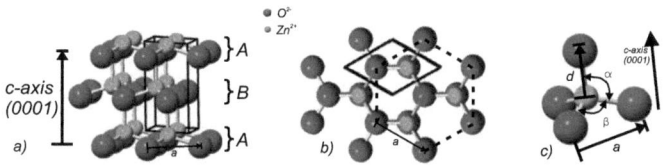

Figure 1.1: Wurtzite crystal structure of ZnO. Each atom is bound tetrahedrally to four atoms of the other kind. In a) the Oxygen and Zinc terminated surfaces are visible. b) visualizes the hexagonal character of the structure (top view of a)). The black box shows the unit cell. The dotted line indicates the hexagon of atoms of the same layer. c) depicts the bonding angels.

benefit from the improvement of growth techniques at the end of the last century. Earlier attempts failed due to the large lattice mismatch of GaN to most of the common substrates (e.g. sapphire). Indirectly this influenced the growth of ZnO since this material was reported of be successfully grown on sapphire when including a GaN buffer layer [40]. 6H-SiC has also proven to be a suitable substrate for the high quality GaN growth, but the high conductivity of typical SiC compromise applications [45].

property	Gallium nitride	Zinc oxide
formula	GaN	ZnO
appearance	yellow/greenish	white/greenish
density	6.15 g/cm^3	5.606 g/cm^3
molar mass	83.73 g/mol	81.41 g/mol
melting point	>2500 °C	1975 °C

Table 1.1: General properties of GaN and ZnO

1.2.1 Wurtzite crystal structure of ZnO

So far ZnO is known in three crystal structures: zincblende, rocksalt and wurtzite [38]. The first two structures are rarely realized since they need either very high

$a(\text{Å})$	$c(\text{Å})$	c/a	u	method
3.2496	5.2042	1.6018	0.3819	XRD
3.2501	5.2071	1.6021	0.3817	XRD
3.286	5.241	1.595	0.383	*ab initio* LCAO
3.2498	5.2066	1.6021		EDXD
3.2475	5.2075	1.6035		XRD
3.2497	5.206	1.602		powder XRD

Table 1.2: Measured and calculated lattice values for the wurtzite structure of ZnO taken from [38] and references therein.

pressure or uncommon substrates. Under ambient conditions the wurtzite phase is energetically favorable. This structure is composed of two interpenetrating hexagonal closed packed sublatices of cation (Zn^{2+}) and anion(O^{2-}) stapled ABAB... along the c axis. Two layers are displaced by $\sqrt{\frac{1}{3}}\, a$ with a the lattice constant of the hexagonal unit cell. Figure 1.1 shows side and top view of a wurtzite crystal. In a) the oxygen and Zinc surfaces become obvious. In b) the hexagonal character of the structure is pronounced. The black prism shown in both graphics represents the unit cell which consists of two atoms of each kind. The symmetry of ZnO is often referred to as C_{6v} [38] whereas it is clear form 1.1 b) that a rotation of $\pi/3$ lead to altered atom positions of the layer above/below, respectively. Due to the tetrahedral bonding configuration, the crystal environment of each individual atom has a C_{3v} symmetry [38, 46, 47].

Figure 1.1 c) shows a close up of the lattice structure with the bonding angles α and β (109.070° in an ideal wurtzite crystal). The dimensionless u-parameter is defined as the nearest neighbor distance d along the c-axis divided by the c-lattice constant:

$$u = \frac{d}{c} = \frac{3}{8} = 0.375 \text{ (in ideal wurtzite structure)} \quad (1.14)$$

Assuming a fixed bond length it is directly correlated to the lattice constants a and c.

$$u = \frac{1}{3}\frac{a^2}{c^2} + \frac{1}{4} \quad (1.15)$$

The real ZnO crystal deviates from the ideal wurtzite structure and the u-parameter is often used to quantify this deviation.

Table 1.2.1 compiles experimental results for the ZnO lattice by XRD and energy

$a(\text{Å})$	$c(\text{Å})$	c/a	u	reference(year)
3.18	5.18	1.62	-	[48] (1938)
3.189	5.185	1.625	-	[45] (1992)
3.28940	5.18614	1.62606	0.3789	[49] (2004)
3.18926	5.18523	1.6258	-	[50] (2007)

Table 1.3: Lattice values for the GaN wurtzite structure. The last digits of the two bottom rows are still under debate [50].

dispersive x-ray diffraction (EDXD) as well as *ab initio* theoretical calculation based on a linear combination of atomic orbitals (LCAO) .

1.2.2 Wurtzite crystal structure of GaN

The second host material for DMS – namely GaN – in this work has the same crystal structure like ZnO. Ga^{3+} and N^{3-} are placed similar to ZnO on tetrahedral coordinated lattice cites for cation and anion, respectively. The bonds in this III-V semiconductor are much more covalent and not predominantly ionic like in II-VI systems.

GaN is also known in Zinc-blende structure. Wurtzite structure and Zincblende structure differ only by the stacking sequence of the layers which is ABA...and of ABCA...for the cubic phase. Table 1.3 gives exemplarily some experimental results for lattice parameters of GaN in wurtzite structure.

1.3 Electron magnetic resonance

A elegant introduction to magnetic resonance is given by Charles P. Poole [51]. Essential prerequisite for this phenomenon is an uncompensated magnetic moment. Starting with the Hamiltonian describing the energy levels of an atom or a radical containing unpaired electrons and nuclei with nonzero spin, one can discuss the different contributions with regard to their importance for magnetic resonance.

$$H = H_{el} + H_{CF} + H_{LS} + H_{SS} + H_{Ze} + H_{Zn} + H_{HF} + H_{II} + H_{Q} \quad (1.16)$$

Which is actually the sum of the contributions from the electron's kinetic and coulomb energy (H_{el}), the crystal field energy (H_{CF}), the spin-orbit coupling (H_{LS}), the spin-spin interaction (H_{SS}), the electron Zeeman energy (H_{Ze}), the nuclear Zeeman energy (H_{Zn}), the hyperfine energy (H_{HF}), the nucleus-nucleus interaction (H_{II}) and the quadrupolar nuclear contribution (H_Q). The first part of equation 1.16 contains the kinetic and potential energy of the electrons relative to the nuclei, including electron-electron repulsion energies. Summing up over the index i and j (electrons) and n (nuclei), H_{el} can be written as:

$$H_{el} = \sum_i \frac{p_i^2}{2m} + \sum_{i,n} \frac{z_n e^2}{r_{ni}} + \sum_{i>j} \frac{e^2}{r_{ij}} \quad (1.17)$$

With p_i = momentum of electron i, z_n=atomic number and e being the electron charge.

The energies derived from equation 1.17 are of the order of 10^4-10^5 cm^{-1} ($\approx$ 1-10 eV) which affects the optical properties of atoms. Typical energies for magnetic resonance measurements - mainly determined by the Zeeman energy - are orders of magnitude lower. Therefore H_{el} plays only a minor role in magnetic resonance. The second sub-Hamiltonian in 1.16 on the contrary is one of most important ones when the magnetic center is incorporated in a solid.

$$H_{CF} = \sum_{i,j} \frac{Q_j}{r_{i,j}} \quad (1.18)$$

This crystal field Hamiltonian describes the surrounding electric potential within a point charge approximation. The summation is over the Q_i ionic charges and the i electrons. It can be interpreted as a spatially anisotropic Stark effect acting on ionic and covalent bonds. This anisotropy is the reason for the high information gain derived from angular dependent measurements.

The spin-orbit interaction is written as:

$$H_{LS} = \lambda \vec{L}\vec{S} \quad (1.19)$$

With λ = spin-orbit coupling constant, $\vec{L}$ and $\vec{S}$ being the orbital and spin angular momentum. L and S are coupled with magnetic moments, therefore the relevance of this Hamiltonian for magnetic resonance is obvious. Nevertheless it plays a twofold role, since its energy contribution is often negligible compared to the crystal field splitting for the first series of transition elements, whereas it becomes predominant for the rare earths.

The fourth sub-Hamiltonian in equation 1.16 describes the spin-spin interaction.

It is rather complicate, since it can have different origins. In case of dominating exchange interaction it can be written as [52]:

$$H_{SS} = \vec{S}_i \, \overleftrightarrow{J}_{i,j} \, \vec{S}_j \tag{1.20}$$

With $S_{i,j}$ = spin of electron i and j and $\overleftrightarrow{J}_{i,j}$ being the spin-spin exchange tensor. Equation 1.20 gives the general expression for an anisotropic exchange interaction of different spins with each other.

A form containing the so called "zero field splitting" parameters D and E is of particular interest for multi-electron systems. E vanishes in case of axial crystal symmetry, therfore it is only of interest in highly asymmetric crystals which will be not subject of this work.

$$H_{SS} = D\left[S_z^2 - \frac{1}{3}S(S+1)\right] + E(S_x^2 - S_y^2) \tag{1.21}$$

The zero field splitting lifts the degeneracy of energy levels and leads to various fine structure lines in magnetic resonance. The energy of the spin-spin interaction is frequently of the same order of magnitude like the Zeeman-energy, which leads to strong angular dependent spectra.

It should be mentioned that the latter three Hamiltonians are strongly correlated. A perfect cubic crystal field for example can lead to a quenched orbital moment with in turn can lead to an isotropic spectrum because of $< L_z >= 0$.

The sensitivity of electron magnetic resonance measurements is typically enhanced by the use of a resonator technique, which will be described in section 2.2. Due to the geometrical conditions of the cavity the feasible frequency range is limited to frequencies close to the eigenfrequency of the resonator. This in turn means that the resonance condition of a paramagnetic center can not be obtained by change of $h\nu$. The Zeeman Hamiltonian offers an easy possibility to 'tune' the resonance condition of probed atoms, ions or molecules - just by applying an external magnetic field [3].

$$H_{Ze} = \mu_B \, \vec{S} \, \overleftrightarrow{g} \, \vec{H} \tag{1.22}$$

With μ_B being the Bohr magneton, $\vec{B}$= external magnetic field, $\vec{S}$ = spin and $\overleftrightarrow{g}$ = g-tensor containing the interaction of $\vec{H}$ with the orbital magnetic moment. In this respect one can call the Zeeman term the most important for magnetic resonance measurements.

---

[3]which is only valid for fields much weaker than the intra-atomic magnetic fields

Analogous to the electronic Zeeman term the nuclear Zeeman term describes the interaction of the magnetic moment of one or more than one nuclei with an external magnetic field.

$$H_{Zn} = -g_n \, \mu_n \, \vec{H} \, \vec{I} \qquad (1.23)$$

With g_n=nuclear g-factor, μ_n= nuclear magneton and $\vec{I}$ being the nuclear spin. Since the nuclear magneton is about three orders of magnitude smaller than the magneton of an electron this term is negligible if one deals with electron resonance. In contrast to the previous the interaction of nuclear moments with an electron can often be detected. Especially for molecules the interaction of an electron with several nuclei is often described in literature and yields valuable information. The corresponding Hamiltonian for this hyperfine interaction can be written as:

$$H_{HF} = \vec{S} \sum_i \overleftrightarrow{A}_i \, \vec{I}_i \qquad (1.24)$$

To conclude this part the last two Hamiltonians have to be shortly addressed. Considering the size of the nuclear magnetic moment it is understandable that

$$H_{II} = \sum_{i>j} \vec{I}_i \, \overleftrightarrow{J}_{i,j} \, \vec{J}_j \qquad (1.25)$$

- the spin-spin interaction for nuclei - is only relevant for nuclear magnetic resonance. The last term in 1.16 appears due to the influence of the quadrupole moment of nuclei. It originates from the inhomogeneous charge distribution of the nucleus which leads to an electrical field which in turn interacts with the electrical field produced by the electrons. The main contribution can be written as:

$$H_Q = \frac{e^2 Q}{4I(2I-1)} \left(\frac{\partial^2 V}{\partial z^2}\right) \left[3I_z^2 - I(I+1) + \eta(I_x^2 - I_y^2)\right] \qquad (1.26)$$

with

$$\eta = \frac{\frac{\partial^2 V}{\partial x^2} - \frac{\partial^2 V}{\partial y^2}}{\frac{\partial^2 V}{\partial z^2}} \qquad (1.27)$$

With Q=quadrupole moment, $\frac{\partial^2 V}{\partial z^2}$ being electric field gradient and η=the asymmetry parameter. It is rather rarely observed in electron spin resonance but plays a more prominent role in nuclear resonance, electron-nuclear double resonance (ENDOR) or Mößbauer spectroscopy.

THE SPLITTING OF Co^{2+}-ION STATES IN ZnO

Energy	[cm^{-1}]	[eV]
H_{el}	$\approx 10^4$-10^5	$\approx 10^0$-10^1
H_{CF}	$\approx 10^0$-10^1	$\approx 10^{-4}$-10^{-3}
H_{SL}	$\approx 10^{-1}$-10^2	$\approx 10^{-5}$-10^{-2}
H_{SS}	$\approx -10^4$	$\approx -10^0$
H_{Ze}	$\approx -10^0$	$\approx -10^{-4}$
H_{Zn}	$\approx -10^{-3}$	$\approx -10^{-7}$
H_{HF}	$\approx -10^0$	$\approx -10^{-4}$
H_{II}	$\approx -10^{-6}$	$\approx -10^{-10}$
H_Q	$\approx -10^{-3}$	$\approx -10^{-7}$

Table 1.4: Overview of energy contributions to magnetic resonance and typical values given in inverse centimeters and electron volt.

1.3.1 The splitting of Co^{2+}-ion states in ZnO

The interaction of the magnetic ions in DMS is supposed be a key factor for developing ferromagnetism. Isolated ions will be referred to as ions without any further interaction than the ones originating from the pure ZnO environment. Therefore this section shows an example of splitting of energy states of isolated ions as it is observed by electron spin resonance (ESR) or electron paramagnetic resonance (EPR), respectively [4]. The term "ideal dilute paramagnet" is often used in literature as description but will be avoided in this work because of the possible confusions with the term "dilute" used for DMS.

When incorporated into the wurtzite lattice of ZnO and substituting the Zn-ion the energy levels of the divalent Co-ion change significantly due to the influence of the surrounding ions reflected by the Hamiltonians mentioned in 1.3. Even though the basic quantum mechanical description of crystal fields dates back to the first part of the last century [53], it is noteworthy to point out that it is still subject to recent research [54]. If one reduces equation 1.16 to the contributions relevant for ESR measurements, one gets:

$$H_S = \mu_B \vec{B} \overleftrightarrow{g} \vec{S} + \vec{S} \overleftrightarrow{A} \vec{I} + D \left[S_z^2 - \frac{1}{3}S(S+1) \right] \tag{1.28}$$

---

[4]Since many resonance spectra will be presented which origin might be ferromagnetic or paramagnetic the term ESR will be preferred.

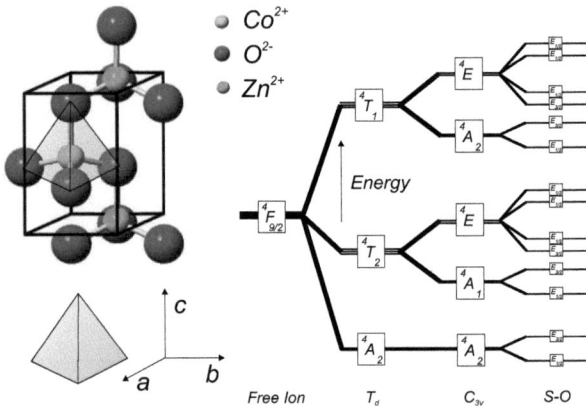

Figure 1.2: Qualitative energy splitting of Co^{2+} in ZnO [54]

Without any external influences like chemical bonds or magnetic or electric fields the Co^{2+}-ion with its [Ar] 3d^7 electron configuration has a $^4F_{\frac{9}{2}}$ ground state. Therefore also in the environment of a crystal there will be $4 \times 7 = 28$ (spin multiplicity × orbital multiplicity) states.

The influence of the crystal field of the wurtzite structure can be split into two contributions. As shown in figure 1.2 the tetrahedral part has a T$_d$ cubic point symmetry which causes a threefold energy state separation [5]. The corresponding ground state belongs to a $S = \frac{3}{2}$, $^4A_2$ two times degenerated multiplet. Additionally two three times degenerated exited states develop, namely $^4T_2$ and $^4T_1$.

As discussed in section 1.2.1 the wurtzite lattice of ZnO does not provide perfect tetrahedral bondings. In terms of crystal fields this results in a additional trigonal component which leads to reduced degeneracy of the exited states.

Finally the spin-orbit coupling causes the energy levels to split into Kramers doublets. The resulting lowest lying E$_{\frac{1}{2}}$ and E$_{\frac{3}{2}}$ states are separated by a 2D zero field splitting. For typical X-band measurements $D >> \hbar\omega$ and thus for Co^{2+} in ZnO only the $m_s = \pm\frac{1}{2}$ transition is observed in ESR measurements.

So far the hyperfine interaction has been neglected. The natural abundance of cobalt is limited to only one isotope, ^{57}Co with a nuclear spin I of $\frac{7}{2}$. The electron's wave

---

[5] neglecting the hyperfine splitting which will be discussed later on.

POWDER SPECTRA

function of the ground state does not vanish at the nucleus, therefore there is a residual probability of the electron to be close enough to interact. This splits the corresponding energy levels into 8 equally separated ones, because of the possible half integer values of I_z from $-\frac{7}{2}$ to $+\frac{7}{2}$. The admixture of s-like wave function components to the ground state is anisotropic and the hyperfine splitting shows a strong angular dependence.

Experimentally obtained spectra and values for $\overleftrightarrow{g}$ and $\overleftrightarrow{A}$ will be given in section 3.2. Note that for the proper determination of D either optical or high field measurements are necessary [55, 56].

1.3.2 Powderspectra

In the following the derivation of the magnetic resonance spectra of powders with randomly oriented nanocrystals will be shortly recapitulated. The paramagnetic centers inside the crystallites are assumed to have a uniaxial anisotropy. This model captures the essence of Co doped ZnO powder.

The angular dependence of the resonance field for a single paramagnetic center with uniaxial anisotropy is given by:

$$B_{res} = \frac{\hbar\omega}{g_{eff}\mu_B} = \frac{1}{\sqrt{g_\parallel^2 \cos(\theta)^2 + g_\perp^2 \sin(\theta)^2}} \quad (1.29)$$

This functional dependency is depicted in figure 1.3 for exemplarily chosen values of $B_{res\parallel}$ and $B_{res\perp}$. Without loss of generality $B_{res\parallel} > B_{res\perp} (g_\parallel < g_\perp)$ is assumed. With increasing difference of the two field values, namely the anisotropy, the peaks of the out-of-plane geometry become pronounced. In the limiting case of $B_{res\perp} = 0$ one would yield:

$$B_{res} = \begin{cases} B_{res\parallel} & \text{in case of } \theta = n \cdot \pi \\ B_{res\perp} = 0 & \text{in case of } \theta \neq n \cdot \pi \end{cases} \quad (1.30)$$

This makes already clear that in case of a integration over a random orientation of the nanoparticles with respect to the applied field only $B_{res\perp}$ will significantly contribute to the signal.

The drawing in figure 1.3 depicts the mathematical considerations for a more quantitative analysis. A random orientation of the paramagnetic center means that the density of magnetic moments on the given sphere should be constant. The fraction of the surface of a differential slice of the sphere with respect to the whole sphere

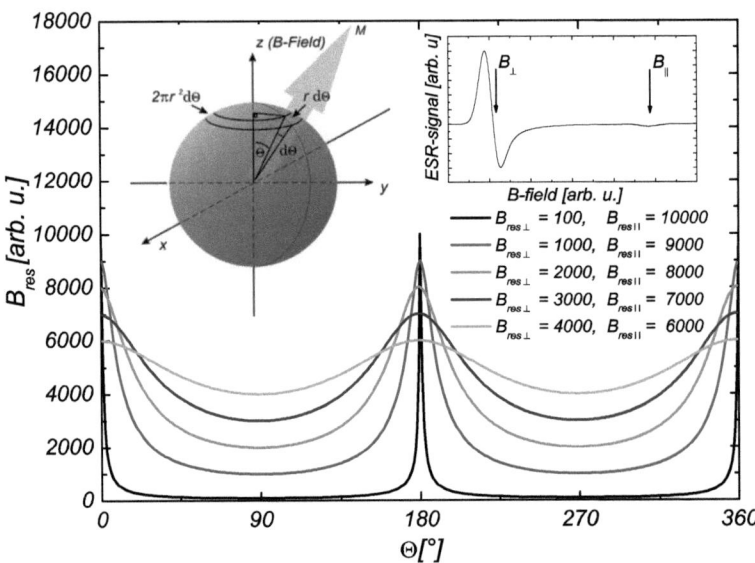

Figure 1.3: Angular dependence of resonance field in case of uniaxial anisotropy. The different pairs of resonance fields (g-values) reflect the different strength of the anisotropy. The inset shows the simulation of a resonance spectrum of a powder with randomly oriented nanocrystals.

surface is given by:

$$d\Omega = \frac{\overbrace{2\pi r^2 sin\theta d\theta}^{\text{slice surface}}}{\underbrace{4\pi r^2}_{\text{sphere surface}}} = \frac{1}{2}sin\theta d\theta \tag{1.31}$$

Since the resonance field of the individual paramagnetic center depends only on the angle θ, the differential probability $p(B)$ to find a paramagnetic center under resonance condition at a field B is proportional to [57]:

$$p(B)dB \propto d\Omega \propto sin\theta d\theta \Leftrightarrow p(B) \propto \frac{sin\theta}{dB/d\theta} \tag{1.32}$$

FERROMAGNETIC RESONANCE

The derivative of equation 1.29 can be inserted:

$$p(B) \propto \frac{\hbar\omega}{\mu_B} \frac{(g_\parallel^2 \cos(\theta)^2 + g_\perp^2 \sin(\theta)^2)^{\frac{3}{2}}}{(g_\perp^2 - g_\parallel^2)\cos(\theta)} \quad (1.33)$$

Substituting the resonance field condition 1.29 again results in:

$$p(B) \propto \left(\frac{\hbar\omega}{\mu_B}\right)^2 \frac{1}{B_{res}^3 (g_\perp^2 - g_\parallel^2)\cos(\theta)} \quad (1.34)$$

For $\theta=\frac{\pi}{2}$ this term rises to infinity whereas for $\theta=0$ it remains finite. For proper calculation of the absorption spectrum of such a system 1.33 has to be integrated and a finite linewidth has to be considered. A detailed description of the solution of the integral and linewidth considerations can be found in [58].

The inset of figure 1.3 shows exemplarily a simulated powder spectrum of Co:ZnO.

1.3.3 Ferromagnetic resonance (FMR)

The introduction of the theory of ferromagnetic resonance (FMR) will remain fragmentary, since the experimental and discussion part will only make sporadic use of it. A detailed description of FMR can be found in the first works of C. Kittel [59, 60]. More recent works with special regard to layered structures can be found for example in [61, 62].

In principle the Hamiltonians introduced in section 1.3 are suitable for most kinds of magnetic resonances of solid state systems [6]. Nevertheless in the case of FMR a slightly altered approach has proven to be more instructive. The main reason for a different description of ferromagnetic resonance is due to the fact that in ferromagnets the magnetic atoms or ions can not be treated separately. To use the term introduced in section 1.3: The magnetic centers are not isolated. Instead a strong exchange coupling leads to a long range magnetic order. In case of an external magnetic field the resulting magnetization starts to precess around the field direction. As mentioned before this case is in principle covered by the spin-spin Hamiltonian. However, a proper description leads to an complex quantum-mechanical many-body problem [63]. Due to the large amount of correlated spins a classical approach is appropriate as well, according to the correspondence principle of quantum mechanics: The exchange coupling of the magnetic moments results in a total magnetization $\vec{M}$, which interacts with the magnetic component of the microwave radiation. This

---
[6] neglecting more exotic cases like cyclotron resonance or Pauli paramagnetism

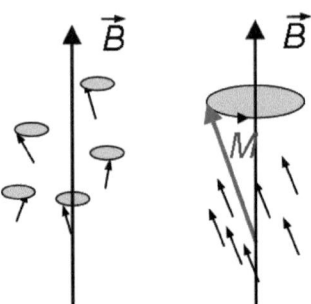

Figure 1.4: Sketch of paramagnetic and ferromagnetic spins. Individual precession of spins around the magnetic field axis in case of paramagnetism (left). Coherent rotation due to exchange coupling in a ferromagnet(right).

causes the magnetization to start to precess around its equilibrium orientation in the external magnetic field. The dynamics of this process is described by the well known Landau-Lifshitz equation:

$$\frac{dM}{dt} = -\gamma \left(\vec{M} \times \vec{B}^{eff} \right) \quad \text{with } \vec{B}^{eff} = \vec{B}^{int} + \vec{B}^{ext} + \vec{B}^{MW} \tag{1.35}$$

If the damping is taken into account the equation is expanded by a term first introduced by Gilbert:

$$\frac{dM}{dt} = -\gamma \left(\vec{M} \times \vec{B}^{eff} \right) + \frac{\eta}{M} \left(\vec{M} \times \frac{dM}{dt} \right) \tag{1.36}$$

With η being the damping parameter. For completeness it has to be mentioned that the perception of a total magnetization and its dynamic is in some respect too simplified. A more accurate description stresses the coherence of the spin precession. The damping effect therefore describes a dephasing of the precession of coupled spins. The additional presence of a perpendicular oriented high frequency magnetic field of a microwave perturbs this spin rotation. This effect is most pronounced in case of resonant absorption, i.e. the frequency matches the Larmor frequency. Respective equations have been first reported by F. Bloch for nuclear magnetic moments [64]. The prove of ferromagnetism solely by magnetic resonance measurements is difficult. The reason are paramagnetic resonance lines which can show similar properties. Probably the best criterion to distinguish both is the temperature dependence. Approaching the Curie temperature of a ferromagnet from lower temperatures the magnetic resonance line is expected to broaden, since the thermal excitation reduces the spin relaxation time because of enhanced phonon scattering. Further the temperature dependent anisotropy contributions lead to a shift of the resonance field towards g = 2 ("$\frac{\omega}{\gamma}$" [7]). Finally, above T_C a ferromagnet becomes a paramagnet and the ferromagnetic resonance signal vanishes.

Paramagnetic resonance signals are generally expected to decrease with increasing temperature like 1/T reflecting the Curie law. Also for paramagnetic ions a line broadening is expected due to phonon interactions.

Unfortunately several physical and technical conditions can further complicate the identification of the origin of a magnetic resonance signal. For instance the change of valency with temperature of a dopant in a semiconductor might mimic a Curie temperature. Therefore magnetic resonance measurements need complementary techniques like SQUID magnetometry to prove the existence of ferromagnetism.

1.3.4 Transition from ESR to FMR

As already mentioned in 1.3.1 to observe single ion paramagnetic spectra it is a decisive prerequisite that these magnetic centers are isolated, i. e. without interaction among each other. On the other hand for FMR exactly this interaction is essential - no matter whether it is direct exchange, superexchange or Ruderman-Kittel-Kasuya-Yosida (RKKY) coupling. Ferromagnetism of DMS systems is sometimes reported even at ultra high dilution of the dopant atom [65]. With respect to potential interactions special care has to be taken concerning the statistics of dopant distribution. R. Behringer published a work already in 1958 where he calculated probabilities of dopant atoms to have a next neighbor or even two next neighbors of the same kind for different host lattices [66]. The calculation for the hcp-lattice can be directly transferred to cobalt atoms in ZnO, if only the cation sublattice is considered [8].

Figure 1.5 shows the likelihood for singles, doubles and triples to occur in the wurtzite structure for varying dopant concentration p. The curves are calculated according to the equations given in table 1.5. The label triple A and B refer to so called open and closed configurations. The latter have a reduced amount of bond-

---
[7] $\gamma = \frac{g_J \mu_B}{\hbar}$ - thus g = 2 is valid for spin-only systems
[8] As will be discussed in chapter 5.2 cobalt is indeed very prone to substitute for zinc.

ings to other atoms and therefore a different probability to occur.

$$\begin{aligned}
\text{singles:} &\quad (1-p)^{12} \\
\text{doubles:} &\quad 12p(1-p)^{18} \\
\text{triples A (open):} &\quad 18p^2(1-p)^{23}[5(1-p)+2] \\
\text{triples B (closed):} &\quad 3p^2(1-p)^{21}[1+6(1-p)+(1-p)^2]
\end{aligned}$$

Table 1.5: Probabilities of dopant configurations depending on the concentration p in a hcp lattice according to [66]

The vertical black line corresponds to a dopant concentration of 1%. At a first glance it is quite counterintuitve that at this relatively low concentration already more 11% of dopant atoms have at least one next cation neighbor which is of the same kind. It becomes more understandable if one takes into account that due to the wurtzite lattice structure each dopant has twelve next cation neighbors.

For a ferromagnetic response of a sample normally several thousands of coupled atoms have to exists. A intermediate range of a few coupled atoms has not been addressed in this chapter so far.

Two important effects have to be taken into account:

i) In case of a coupling between the electrons of atoms governed by the exchange interaction a new energy level scheme of the pair, trimer or more atoms containing structure will develop. A Hamiltonian similar to 1.20 can be applied for the exchange interaction of electrons of different atoms leading to respective multiplets. Depending on the strength of the interaction weak, intermediate and strong coupling has to be distinguished. The latter results in different commuting operators and therefore energy schemes. Consequently magnetic resonance spectra are likely to consist of completely different resonance fields than the single ion spectra. A thorough treatise of exchange coupled systems of oligonuclear systems can be found in [67].

ii) The second effect which has to be considered is the perturbation of the single ion spectra. In magnetic resonance measurements this is reflected by a change of the linewidth of a certain transition. The linewidth in turn is related to the relaxation processes which depopulate the excited energy level.

In general two relaxation times are considered. A fundamental limit of the linewidth is given by the Heisenberg uncertainty relation in the form $\Delta E \Delta t \geq h$ [68]. The respective time is referred to as T_1. The line broadening induced by other mechanisms

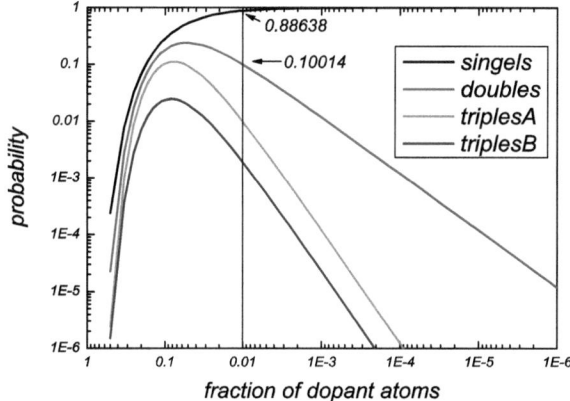

Figure 1.5: Logarithmic plot of probabilities of dopant constellations in wurtzite according to Behringer's equations.

like spin-spin dipolar interactions [69] or spin-nuclear spin interactions are compiled in a time constant T_2. In many cases the linewidth is governed by T_2.

The influences on the linewidth of a resonance discussed so far are referred to as homogeneous broadening. Besides this inhomogeneous broadening may occur. As an example of this effect inhomogeneous internal fields which cause different fractions of a sample to have shifted resonance fields might be considered. In total the line shape represents typically a Gaussian as envelop function of several resonance contributions. Even though the study of magnetic resonance has started in the middle of the last century the precise calculation of resonance linewidth is subject of recent research [70], since it contains valuable information about the interactions of the magnetic centers.

1.4 Element specific investigation methods

Within this work synchrotron radiation was used to study structure and magnetism of DMS with element specificity. Two effects were utilized, namely the X-ray linear dichroism (XLD) and the X-ray magnetic circular dichroism (XMCD). After a short introduction to X-ray absorption spectroscopy (XAS) both effects will be described briefly.

1.4.1 X-ray absorption spectroscopy (XAS)

If light penetrates matter it experiences a exponential decay of the initial intensity I_0. The respective relation is known as Lambert-Beer law:

$$I(x) = I_0 \exp^{-\mu(E)x} \qquad (1.37)$$

$\mu(E)$ is the attenuation coefficient and x the thickness of the material. The attenuation coefficient typically decreases with increasing energy of the radiation. This behavior is mainly caused by the decline of the cross-sections of photo and Compton effect in typical energy regimes. Depending on the atomic composition of the material sudden peaks superimpose absorption spectra. A so called absorption edge is determined by the energy difference of an occupied (typically core level for X-rays) and an unoccupied electron state of an atom. If the incoming photon matches this energy the absorption cross-section sharply increases and the absorption process becomes resonant. These energy values are element specific, which makes X-ray absorption spectroscopy (XAS) (by use of monochromators) an element specific spectroscopic investigation method. The quantum mechanical description of the

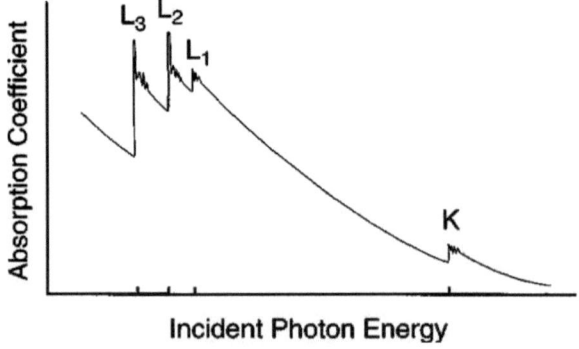

Figure 1.6: Energy dependence of X-ray absorption. Four X-ray absorption edges are shown: K, L_1, L_2, and L_3. (Adapted from [71])

ELEMENT SPECIFIC INVESTIGATION METHODS

absorption process is done by time dependent perturbation theory.

$$H(t) = H_0 + V(t) \tag{1.38}$$

With H_0 being the unperturbed system and $V(t) = \Theta\left(\hat{V}e^{-i\omega t} - \hat{V}e^{i\omega t}\right)$. Assuming a small perturbation the transition probabilities from a initial state (i) to a final state (f) Γ_{if} are given by Fermi's *golden rule*:

$$\Gamma_{if} = \frac{2\pi}{\hbar}\rho\overline{|<f|\hat{V}|i>|}^2 \tag{1.39}$$

ρ represents the density of the final states. Therefore XAS probes the local environment of the atom, since this density of final states for solids strongly depends on crystal structure, hybridization and lattice constants.

Depending on the extent of the energy range of the spectra after the actual absorption edge one distinguishes two types of XAS :

In XANES (also called near edge absorption fine structure (NEXAFS)) the valence state of the selected type of the atom in the sample and the local symmetry of its unoccupied orbitals can be deduced from the shape and energy position of the X-ray absorption edge. In extended x-ray absorption fine structure (EXAFS) interference of the outgoing μ of the electron (-wave) is scattered back in the close environment leading to modulation of μ due to constructive or destructive interference of the outgoing and backscattered wave, which reveal e. g. bonding lengths. There is no strict rule to separate the EXAFS and XANES regime. Typically 50 to 70 eV beyond the absorption edge are used for XANES whereas the EXAFS extends to some hundreds eV.

The straightforward way to detect XAS is to measure the photons transmitted through a sample. Due to the exponential decay of the X-ray beam intensity inside the material this enforces very thin samples and therefore experimentally unhandy conditions like difficult mounting.

Two other types of XAS-detection have been established. Both methods are based on secondary processes that result from the refilling of the empty core state by other electrons. The excess energy of the electron can be emitted via the emission of a photon or an Auger-electron. In *total fluorescence yield* the amount of emitted photons is measured whereas in *total electron yield* the outgoing electrons are detected. The proportionality between these and the X-ray absorption are demonstrated for example in [28, 72, 73].

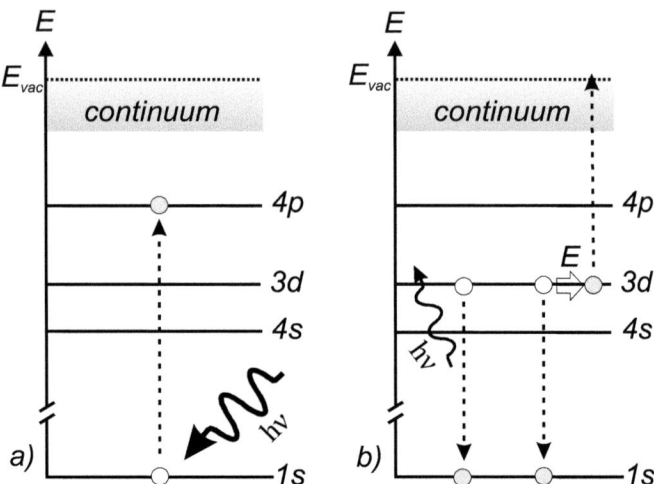

Figure 1.7: XAS at the K-edge of a 3d-transition element in a solid. a) a photoelectron is excited to the valence band. b) two possible relaxation mechanisms: On the left: filling of the empty core state under emission of a photon. On the right: filling of the empty core state under energy transfer to another electron which can leave the atomic potential (Auger-process).

1.4.2 X-ray linear dichroism (XLD)

As mentioned before the XAS probes the density of empty (valence-) states of an atom. Due to the geometry of the bonds to the next neighbor atoms, an anisotropy of the local distribution of the density of empty states is possible. Therefore the lattice structure determines the spatial distribution of the density of empty states capable of receiving a photoelelectron.

This anisotropy can be examined by using linearly polarized X-rays. For crystals with low symmetry, like the wurtzite structure, there will be a different absorption spectrum depending on the polarization direction [9]. The difference of both spectra is called XLD. The examination of the "empty electron states landscape" by the

---

[9]This assumes a respective alignment, i. e. for a wurtzite structure perpendicular and parallel to the c-axis. Since for a uniaxial aligned system the intensity varies like $I(\theta) = 3 <I> cos^2\theta$ (with $<I>$ being the angular averaged intensity) [28] one can easily see that the effect vanishes for example at $\theta = \pm 45°$.

electric field vector is sometimes called "search light effect".
It should be added to this simplified description of XLD, that the given arguments hold only if a spherical symmetric core state, namely an s-state, is assumed. For p-states one would have to average over all ($l = -1, 0, 1 \rightarrow p_{\frac{3}{2}}, p_{\frac{1}{2}}$) states, since only the sum of the orbitals is spherically symmetric.

1.4.3 X-ray magnetic circular dichroism (XMCD)

In case of a circular polarization of the synchrotron light the photons contain an angular momentum - the helicity. It is aligned parallel or antiparallel to the direction of propagation of the photon. The XMCD is yielded by the difference of the absorption curves for both helicities. The XMCD effect is commonly described by a two step model [28], which distinguishes the generation of a spin and/or orbitally polarized photoelectron and the transfer of this photoelectron to free electron state of higher energy. The most frequently used XMCD effect for 3d transition elements is the L_3-edge dichroism. In this case a $2p_{3/2}$ electron is removed by the incident photon. The angular momentum of the photon has to be conserved by transferring it to the absorbing electron. If the absorbing electron state is spin-orbit split (like the $2p_{3/2}$ states) the helicity can be – due to the spin-orbit coupling – partly transferred to the orbital and spin momentum of the photoelectron. Thus spin polarized photoelectrons can be generated by circular polarized photons.

In a second step the exchange split 3d electron states act as a spin dependent detector for the photoelectrons. Unequal amounts of empty spin up and spin down states result in the different absorption spectra resulting in the XMCD effect.

For completeness it should be mentioned that the high information gain from L-edge XMCD is closely related to the so-called sumrules derived for charge, spin and orbital magnetic moment. These have not been used for this work and they will not be treated here. The reader is referred to respective literature [28, 74, 75, 76]. For this work mainly K-edge XMCD spectra have been evaluated. In this case the s-state of the core electron cannot be spin-orbit split, since l = 0. However, a small Zeeman or exchange splitting might be present. The angular momentum of the photon is transferred to the photoelectron solely as orbital momentum. If the final states are spin-orbit split absorption will depend on the helicity of the incoming photon, resulting in a XMCD effect which probes the orbital contribution of the splitting of

28　　　　　　　　　　　　　　　　　　　　　　　THEORETICAL BACKGROUND

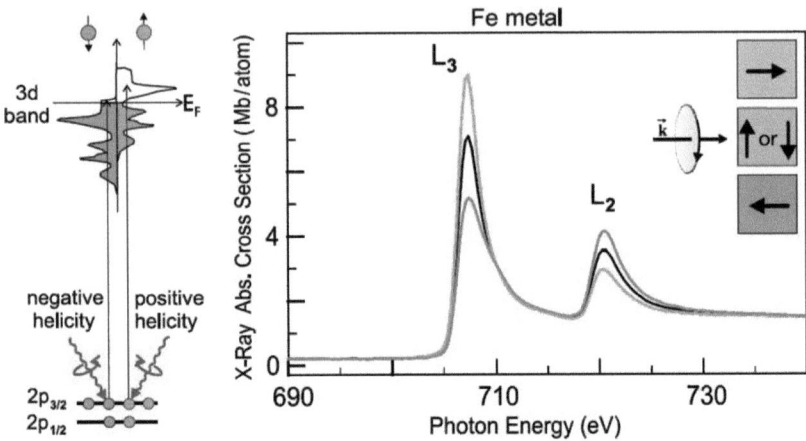

Figure 1.8: XMCD at the Fe L-edge (taken from [28]). Left: Spin polarized photoelectrons are generated depending on the helicity of the photons. The absorption cross-section depends on the spin dependent density of empty states above the Fermi energy. Right: Absorption spectra taken at the Fe L_3 and L_2-edge.

the final states.

Chapter 2

Experimental techniques

The previous chapter has given a short theoretical introduction to the most important techniques used in this work. This chapter will focus on technical aspects and evaluation of the data. The main principles of the experimental techniques will be described with special regard to what will be referred to in the "Experimental results"-chapters.

A thorough introduction of theses techniques is beyond the scope of this work. The given references provide a comprehensive treatise.

2.1 Growth of ZnO and GaN based DMS

The DMS Co:ZnO discussed in this work was predominantly grown by reactive magnetron sputtering. The UHV preparation system offers two possible growth modes. The first one is to sputter from Zn/Co composite targets with a fixed composition. The second one is the simultaneous use of cluster magnetrons which offers varying compositions of host and dopant material. Some details of the preparation chamber are given in appendix C.

Alternative growth techniques for Co:ZnO as pulsed laser deposition (PLD), focused-ion-beam (FIB) implantation and chemical vapor synthesis (CVS) were conducted at collaborating institutions.

Gd:GaN samples were predominantly grown by molecular beam epitaxy (MBE) and provided by collaborators. Low dopant concentrations have been produced by FIB implantation of MBE grown GaN layers.

2.1.1 Reactive magnetron sputtering

$Zn_{1-x}Co_xO$ thin films were deposited on both sides polished 10 mm x 10 mm x 0.5 mm c-plane Al_2O_3 (0001) sapphire single crystal substrates by reactive magnetron sputtering using a metallic ZnCo (nominally 5 % or 10 % Co) target.

To avoid ferromagnetic contamination on the as-prepared substrates, a thorough cleaning procedure for all substrates with acetone, ethanol, deionized water in an ultrasonic bath was performed for 10 minutes, each. For one series of sapphire substrates an additional standard cleaning procedure was used: The substrates were heated up to 1000 °C under UHV conditions [1]. Both gases were additionally purified before entering the chamber. The base pressure of the preparation chamber is in the 10^{-9} mbar range. The working pressure in the chamber during film deposition was controlled at $4 \cdot 10^{-3}$ mbar by adjusting the pumping speed via a throttle valve. The ratio of Ar and O_2 flows was adjusted via separated mass flow controllers. A ratio of 10:1 has been established as optimal composition. Prior to the deposition, the growth rate was measured by a quartz crystal balance placed at the preparation spot. The sputtering powers are typically chosen in the range from 20 W to 30 W. The substrate temperature was controlled at 350 °C or 450 °C during the preparation procedure, respectively. Two independently calibrated thermocouples, placed directly at the heater and close to the sample holder, were used. Details of the preparation chamber and the heater system are given in appendix C and D, respectively.

2.1.2 Additional growth techniques

DMS samples from cooperations with other groups were available which were grown by alternative techniques. Details of the cooperations will be given at the beginning of chapters 4 and 5. The aim of this section is to shortly summarize the different growth techniques.

Pulsed laser deposition

High power short laser pulses were used for plasma ablation from mixed CoO and ZnO powder [77]. $Co_{0.1}Zn_{0.9}O(0001)$ epitaxial films were grown on epiready c-sapphire substrates at a substrate temperature of 550 °C and an O_2 pressure of

---

[1]Later this cleaning procedure turned out to be not necessary.

10 mTorr. A KrF laser (248 nm) was utilized for ablation. The laser repetition rate and power level were 1 Hz and (314 ± 5) mJ/pulse, respectively, yielding 2.4 J/cm^2 incident on the target, and the growth rate was 0.25 Å/sec. For one sample the thickness and Co content was exemplarily cross checked by X-ray reflectometry and proton induced x-ray emission (PIXE) yielding 105 nm and 10.8%, respectively. XRD indicates high structural perfection with a full width at half maximum (FWHM) of 0.38° in ω-rocking curves of the (0002) reflection of ZnO which itself has a FWHM of less than 0.15° in the ω-2θ scan.

Chemical vapor synthesis

Nanocrystalline Co doped powders were synthesized by the chemical reaction in the gas phase using a method known as Chemical Vapor Synthesis (CVS). Stoichiometric mixture of solid, anhydrous Zn(II) acetate (purity 99.995 %, Sigma-Aldrich) and Co(II) acetate (purity 99.99 %, Sigma-Aldrich) precursors was evaporated inside of flash evaporator using a CO_2 laser [78]. The vapor was carried to the hot-wall reactor using He as a carrier gas. In the hot zone of the reactor the precursor vapor decomposes and reacts with oxygen to form oxide particles. The particles are then transported by the gas stream to the particle collector, where they are separated from the gas stream by thermophoresis. The process temperature and pressure were held constant at 1100°C and 20 mbar, respectively.

Molecular beam epitaxy (MBE)

GaN and Gd:GaN samples were provided from four different groups. All GaN and Gd:GaN samples were grown by MBE but methods differ concerning the reactant. Either the growth was Nitrogen plasma assisted or ammonia gas was provided as reactant. Gd was typically provided by co-evaporation [79, 80]. The GaN layers were grown in most cases directly on 6H-SiC(0001) or on sapphire (Al_2O_3). For the latter on some samples buffer layers of GaN or Si:GaN were grown to address the lattice mismatch.

Focus ion beam implantation (FIB)

Gd^{3+} and Co^+ have been implanted into GaN and ZnO, respectively. Implantation energies of 300 keV (Gd^{3+}) and 100 keV (Co^+) resulted in layer thicknesses of 100 nm, according to the simulated penetration profiles using the SRIM simulation software [81]. The implantations were conducted at room temperature; therefore hardly any annealing of implantation damages is expected.

2.2 Setup for electron magnetic resonance measurements

During this work three different microwave (MW) spectrometer to measure ESR were used, predominantly at X-band frequency ($\approx$ 10 GHz). In the following the basic principles of these spectrometers will be described.

All cavity (resonator) bound spectrometer can be interpreted as interference setups - analogous to experiments known from optics. Two signals originating from the same source follow different paths: a signal branch which is guided to the cavity and interacts with the sample and a reference branch which remains unchanged. Both signals are coupled afterwards and detected by a diode.

If the relative phase of both signals is adjusted for example destructive ($\Delta \phi = \pi/2$) a slight phase shift in one of the paths will immediately lead to a resulting signal unequal to zero. In cavity related MW-spectroscopy this phase shift is caused by an altered absorption of the sample inside the cavity.

The left side of Figure 2.1 shows a sketch of an ESR spectrometer. A Gunn-diode [2] is used to produce microwave radiation which is then coupled into a waveguide system. The wave is split into the signal- and the reference branch, respectively. The signal branch is guided to the cavity via a circulator [3] which forwards the reflected signal form the cavity to the detector diode where it interferes with the reference signal.

The microwave in the reference branch is modified in two ways. Its phase is tuned with respect to the cavity reflected wave. The power is adjusted to put an offset power to the detection diode which warrants operation in the optimal regime of the diodes characteristic curve.

---

[2]The microwave radiation is caused by the Gunn-effect which can be understood as cascade effect in a periodically n-type doped semiconductor [82]. Another way of producing MW-radiation is by so called Klystrons which uses free electron beams to generate MW output.

[3]Circulators use the change of polarization induced by ferites to forward an incoming signal only to the next port. If one port is connected to a terminating load they act as isolators (see for example [83]).

SETUP FOR ELECTRON MAGNETIC RESONANCE MEASUREMENTS

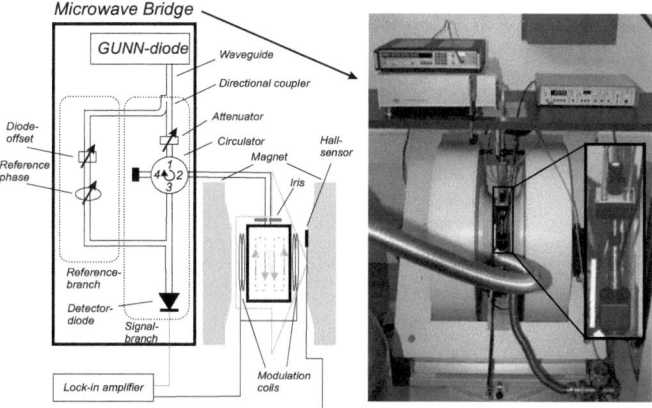

Figure 2.1: Schematic sketch of a ESR spectrometer and photograph of the Bruker XEPR spectrometer. The photograph shows the magnet and microwave bridge with additional frequency counter (top left), low temperature facilities like temperature controller (top right), Helium transfer tube (center) and cryostat (hardy visible behind the transfer tube). The inset shows an enlarged view of the cavity and goniometer.

The cavity is installed inside the gap of an electromagnet setup, which typically ramps up the field during a measurement up to 1.2 T (2.2 T with additional pole shoes). If, due to the magnetic field, a sample gets into its resonance condition the power losses of the cavity are altered, which leads to a change of the resonance frequency of the cavity (+sample).
Two more technical aspects should be mentioned:
The resonance frequency of the cavity changes slightly during the measurement and an automatic frequency control (AFC) is used to keep the system at resonance condition.
To further improve the sensitivity of the measurement a modulated magnetic field is used induced by coils typically installed in the resonator walls (as depicted in violet in figure 2.1 left). A lock-in amplifier compares the modulation and the diode signal - acting as a narrow bandpass filter. The magnetic field is also the abscissa of the measurement; therefore the modulation of the field leads to the detection of the first derivative of the absorption signal. The absorption curve is derived by numerical

integration of the signal.

Cavity enhanced sensitivity

Due to the MW, a standing wave develops inside the resonator. The geometry of the cavity determines the resonance frequency and the possible modes. The magnetic- and electric field inside a cylindrical cavity (which was most frequently used) is exemplarily depicted in Figure 2.1 as green- and yellow arrows, respectively. In this so called TE011-cavity mode [4] the magnetic field of the microwave extends along the z-axis and the electrical field is radial symmetric around this axis. The electrical field has a node at the center of the cavity, thus only the magnetic field is present at the sample position. Due to its axial symmetry the TE011 mode is advantageous for angular dependent measurements.

The ideal coupling between waveguide and the mode inside the cavity is achieved by an iris. The smaller the intrinsic damping the less power has to be injected. To quantify this, the quality factor is defined:

$$Q = \frac{\text{energy stored in cavity}}{\text{energy dissipated per cycle}} = \frac{\omega E_r}{P_{cycle}} \qquad (2.1)$$

The energy stored in the cavity under resonance condition can be written as:

$$E_r = \frac{1}{8\pi\mu_0} \int_{V(cavity)} b_{MW}^2 dV \qquad (2.2)$$

With b_{MW} being the magnetic field inside the cavity.

For magnetic resonance measurements a small sample is mounted inside the resonator. The size of the sample is especially relevant for conducting samples. If the sample significantly extend into regions of the electric field inside the resonator, short-circuits will perturb the mode, which easily can lead to a breakdown of the standing wave inside the resonator.

The MW-power absorbed (per cycle) by the sample is given by:

$$P_{sample} = \frac{1}{2\mu_0} \omega \chi''(\omega) \int_{V(sample)} b_{MW}^2 dV \qquad (2.3)$$

Where χ'' describes the imaginary part of the high frequency susceptibility of the sample, which is defined by:

$$\chi(\omega) = \chi'(\omega) + i\chi''(\omega) \qquad (2.4)$$

---
[4]Transversal electrical mode [84, 85]

SETUP FOR ELECTRON MAGNETIC RESONANCE MEASUREMENTS 35

$\chi'(\omega)$ describes the dispersion effects which can be neglected for alternating electromagnetic fields of high frequency [84]. As stated before the frequency of the experiment is limited to a small range due to the geometrical conditions of the resonator. Subsequently $\chi(\omega) \approx$ const. can be assumed [5]. If the sample itself is not in resonance the additional losses can be neglected. This situation changes if by applying an outer magnetic field a resonant absorption of the magnetic field of the microwave can be achieved. The relative change in the quality factor can be written as:

$$\frac{\Delta Q}{Q_0} = \frac{Q_0 - Q_S}{Q_0} = 1 - \frac{\omega E_r}{P_0 + P_S} \frac{P_0}{\omega E_r} = \frac{P_S}{P_0 + P_S} \quad (2.5)$$

With P_S being the additional power loss due to the sample absorption at its resonance condition and P_0 being the intrinsic loss of the cavity at its resonance condition. It is reasonable to assume that the resonant absorption of the sample only causes a minor additional power loss $\Rightarrow P_0 \gg P_S$, since the resonance condition of the cavity has to be maintained.

$$\frac{P_S}{P_0 + P_S} \approx \frac{P_S}{P_0} \quad (2.6)$$

By use of equations 2.1, 2.2 and 2.3 this can be written as:

$$\frac{\Delta Q}{Q_0} \approx 4\pi \chi'' Q_0 \frac{\int_{V(sample)} b_{MW}^2 dV}{\int_{V(cavity)} b_{MW}^2 dV} \quad (2.7)$$

Q_0 can easily be of the order of 10^3 to 10^4 which is the reason for the high sensitivity of cavity bound ESR setups.

2.2.1 Evaluation of ESR spectra

The line shape of the absorption curve contains information about the microscopic origin of it. Figure 2.2 shows the two most frequently encountered absorption curves, Lorentzian and Gaussian and their first derivative, respectively.
In the following the most important features and conclusions which can be inferred from these line shapes will be summarized. Lorentzian like spectra [6] are typical

---

[5]The resonant absorption of the sample can also be investigated by changing the frequency. This would require a steady change of the cavities geometry, therefore it is rarely realized.

[6]This line shape can be derived from equations first reported for nuclear magnetic moments by F. Bloch [64].

for highly dilut paramagnetic systems whereas Gaussian are rather typical for a convolution of lines.

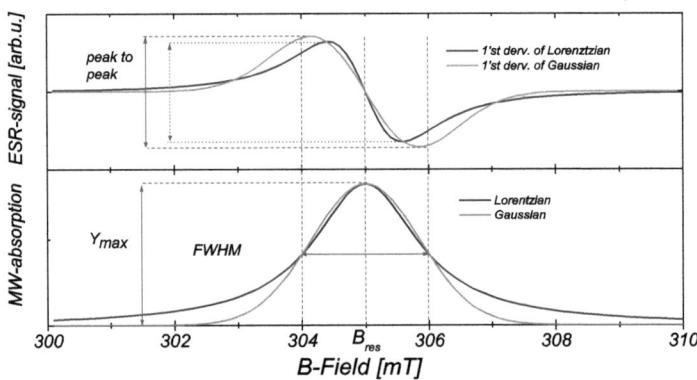

Figure 2.2: Derivatives of Lorentzian and Gaussian line shapes (upper curves) and corresponding integrated signals (lower curves). Both absorption spectra were set to have the same FWHM, B_{res} and Y_{max} to visualize the effect of the absorption line shape on the derivative which is typically measured.

Width

As depicted in figure 2.2 the width of an absorption curve is referred to as the difference of the magnetic field values where the absorption is at half of the maximum value – the full width at half maximum (FWHM).
Generally one has to distinguish between homogeneous and inhomogeneous broadening of a resonance signal. The latter is referred to as resulting from superposition of signals. As an example of inhomogeneous broadening one can consider the mosaicity of polycrystalline films. A slight misalignment of the crystallites with respect to each other will result in a distribution of resonance fields. The following will be limited to effects related to homogeneous broadening.
The fundamental reason for a finite linewidth is the uncertainty principle for energy

SETUP FOR ELECTRON MAGNETIC RESONANCE MEASUREMENTS   37

and time $\Delta E \Delta t \geq h$ [68]. For magnetic resonance measurements this states a relation between the life time of an energy state and its energy value. The resulting linewidth is called natural linewidth and typically referred to by the time constant T_1. The broadening of a resonance line is in most cases not governed by the natural linewidth but by other mechanisms leading to an increased width of the resonance. Paramagnetic centers inside solids are exposed of a variety of interactions. Dipolar fields of other centers can easily cause an significant change in the effective field which leads to resonance field shifts. As a coarse estimation the field of an unpaired electron in a distance of a typical lattice spacing of about 2Å can be calculated:

$$B(r) \propto \pm \frac{2\mu_e}{r^3} \Rightarrow r = 2\text{Å} \Rightarrow B \approx 2\,\text{Gauss} \qquad (2.8)$$

The $\pm$ refers to an arbitrary orientation of the dipole which then in turn would result in a linewidth of 4 Gauss.

In a common solid this simple case has to be expanded by integration over all dipolar moments of the sample. Therefore the perturbation of the single ion has to increase with concentration of the dopant. The respective effect is experimentally frequently reported and will be also shown in section 5.2. Typically perturbations of the single ion condition tend to induce a broadening of the line because of a reduced lifetime of the respective energy states. These effects are referred to by a time constant T_2. A very general treatment of line broadening due to dipolar coupling can be found in [69].

In the same work the effect of exchange interaction between paramagnetic ions is discussed. A additional contribution to the fourth moment of the frequency is shown to be increased, which causes the resonance line to peak stronger. This effect of a reduced linewidth is called exchange narrowing. This must be not mistaken with the case of dominating exchange interaction if a material becomes (anti-)ferromagnetic and resonance conditions are governed by different effects as mentioned in section 1.3.3. In general it is rather difficult to identify the microscopic origin of line broadening/narrowing due to the variety of possible mechanisms involved.

Intensity

As can be seen from equation 2.3 the absorption is direct proportional to the imaginary part of the high frequency susceptibility. The magnetization M can be derived

by integration from magnetic resonance measurements, since $\chi = \frac{dM}{dH}$ [7].
A quantitative determination of M from ESR spectra is only possible with reference samples of known magnetization. Two types of reference samples were used for this work. Most measurements were conducted with $CuSO_4 \cdot 5H_2O$ crystals fixed on a sapphire substrate. Copper(II) sulfate pentahydrat is one of the first and most studied substances by ESR [86]. Diphenylpicrylhydrazyl (DPPH) was rarely used. Its small linewidth and its g-value of 2.0032 makes it a valuable tool for field calibration.

Each sample in a resonator will cause a different damping which leads subsequently to altered sensitivity according to equation 2.7. Besides this many technical settings have to be considered like modulation amplitude and frequency, time constant, gain and conversion time.

Resonance field

The by far most important information is gained from the magnetic field value of the resonance. It can be easily identified as the zero crossing point of the first derivative of the absorption signal. The resonance condition

$$E_i - E_j = h\nu = g\mu_B B \tag{2.9}$$

directly yields information about the energy splitting of two electron states E_i, E_j. In general g is a tensor and angular dependent measurements are required. For a wurtzite structure the g-tensor can be simplified with spherical coordinates for in-plane- to out-of-plane measurements (see for example [57]):

$$\overleftrightarrow{g_{ij}} = \begin{bmatrix} g_{xx} & g_{xy} & g_{xz} \\ g_{yx} & g_{yy} & g_{yz} \\ g_{zx} & g_{zy} & g_{zz} \end{bmatrix} \overset{\text{rotation in xz or yz}}{\Rightarrow} \quad g = \sqrt{g_\parallel^2 \cos(\theta)^2 + g_\perp^2 \sin(\theta)^2} \tag{2.10}$$

Effective g-parameters, which can be experimentally accessed, are commonly used, since the microscopic symmetry of a paramagnetic center might not be determined from the macroscopic appearance of the sample. Even though g_{eff} has proven to be a successful description of experiments, it comprises several influences on the resonance field like e.g. inner fields, spin-orbit coupling and zero field splitting. The

---
[7] Assuming a known sample volume, since $M = \frac{m}{V}$

angular dependency of highly dilute paramagnetic resonances reflect the crystal symmetry [8]. In turn lattice parameter can be derived from resonance conditions. The shift of resonance fields can be understood within the framework of crystal/ligand-field theory as shown for example in 1.3.1. The coupling to the crystal field is provided by the orbital moment, therefore the spectra also contain information about the spin-orbit coupling. In addition the number of fine structure resonance lines of a paramagnetic center can be often used to identify the valence state. Including the hyperfine interaction the resonance fields can also give information about the nucleus as e.g. the nuclear magnetic moment.

2.3 Superconducting quantum interference device (SQUID)

SQUID measurements are nowadays a standard characterization technique for magnetic properties. Throughout this work a MPMS Quantum Design magnetometer was used. This section will not deal with the fundamental function principles of the SQUID sensor itself but with certain problems arising if very small ferromagnetic signals on a diamagnetic background should be sensed. Different from the rest of this work this section will use the unit *emu* (electro magnetic unit), which is the directly provided unit by the magnetometer. Conversion can be easily calculated by:

$$1\ emu = 10^{-3} Am^2 \tag{2.11}$$

In the manufactures MPMS Application Note 1014-213 it is written: *"The MPMS is an extremely sensitive instrument. However, contributions from multiple sources can cause limitations in achieving the full potential of this sensitivity."* This is in particular valid for measurements of DMS samples as will be shown in the following and is reported in [87].

For all measurements the RSO transport was used due to its higher sensitivity compared to the dc-transport mode. Samples were mounted in an clear drinking straw which is recommended by Quantum Design to yield minimum background. Sample pieces were cleaved such that they could be fixed between the walls of the straw

---

[8]This might be not true for ferromagnetic resonance signals since other anisotropy contributions like shape anisotropy come into play.

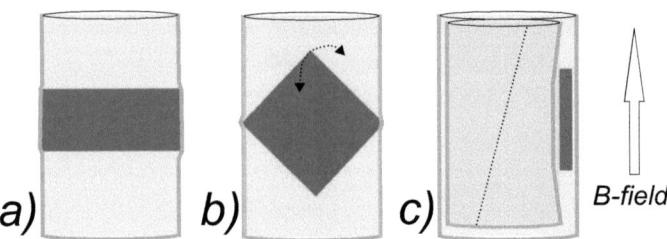

Figure 2.3: Sketch of different fixing of samples inside the straw

(see figure 2.3 a)). For in-plane/out-of-plane measurements quadratic samples were chosen, which were fixed at two diagonal corners. This mounting enables easy rotation to an out-of-plane geometry (figure 2.3 b)). Note that this fixing is crucial if in- and out-of-plane measurements should be comparable, since the background remains identical. Small sample pieces had to be clamped in between the wall of an additional straw and the main straw. The inner straw has to be cut lengthwise and twisted to be insertable (figure 2.3 c)).

Figure 2.4 presents ferromagnetic like signatures in hysteresis measurements of sapphire after cleaving. The raw data show a clear discontinuity of the signal after passing the zero value. Subtraction of the diamagnetic background [9] yields a step-like behavior with a saturation magnetization as shown in b). In c) the corresponding x-position [10] of the measured magnetic moment is plotted. A big step is correlated with the kinks in the hysteresis curves in a) and b). Finally the same sample was measured after intensive cleaning. The respective curves (open symbols) demonstrate that the previous signal may stem from ferromagnetic contaminations. This signature of the x-position shown in c) is typical for edge contamination. During a measurement the sample is moved along the pick up coils (x-pos.) while the respective voltage is recorded. The iterative mode of the SQUID uses a fitting routine which accounts for possible position shifts, for example due to temperature changes, by fitting both amplitude and position of the signal. Figure 2.5 explains the observed behavior. The plots a) – d) depict approximated SQUID voltage curves

---

[9]The diamagnetic background is derived by assuming a linear response for high fields and high temperature.

[10]This nomenclature is adopted from Quantum Design. In fact, the sample moves up/downwards, i.e. the B-field direction which is typically referred to as z-position.

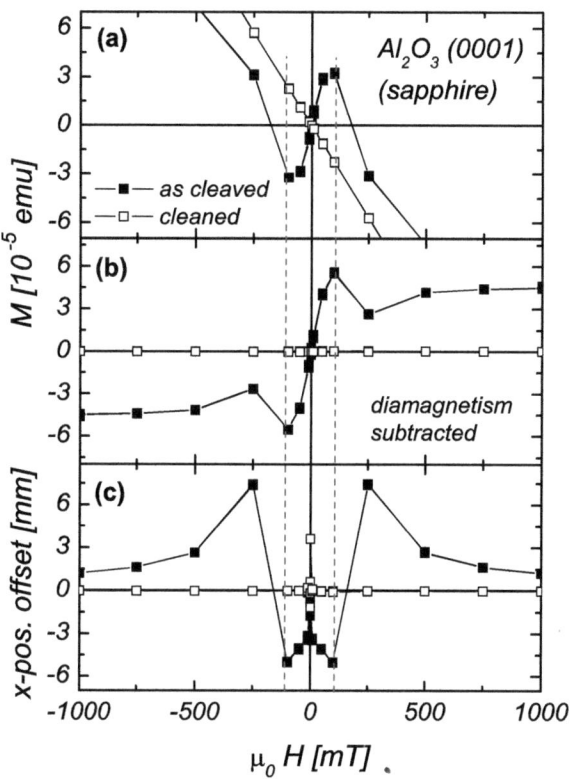

Figure 2.4: SQUID hysteresis measurements of a cleaved sapphire substrate before and after cleaning. In a) the raw data are shown. After subtraction of the diamagnetic background a step like behavior of the uncleaned sample is visible. c) shows the offset of the x-position. For low fields the center of the magnetization is obviously shifted towards the sample edge for the contaminated sample. The red dotted lines indicate the change of the governing magnetic contribution.

or magnetization values, respectively, corresponding to the sample position [11]. The blue dotted line mimics the diamagnetic signal of the sample, which naturally originates from the center of gravity of the substrate. The green dotted line mimics a ferromagnetic contamination at the edge of the sample. The bold red line presents the superposition.

Depending on the applied field the contributions of the respective fractions vary. Whereas in a) the signal is governed by the diamagnetic signal, a clear shift of the peak position is already caused in b). With increasing ferromagnetic contribution the displacement from the actual center of the sample increases (black arrows). In d) two peaks of the same size but opposite sign and far separated are visible. Further increase of the weight of the ferromagnetic contribution will make the fitting routine to change to the peak position which implies a change of the magnetization sign. This is exactly what is observed in 2.4 c).

This relatively easy identification of contamination is only possible in case of upper or lower contaminations, since a centered contribution will not cause the described signatures. Nevertheless it turned out to be good practice to simultaneously check the x-position during measurements, since many contaminations originate from improper cleaving.

Besides a possible contamination of a sample the SQUID itself might cause artifacts which might lead to the attribution of wrong magnetic properties to the sample. As shown in more detailed in [87] the magnetic field control of the SQUID can cause offset fields which lead to inverted hysteresis curves or FC/ZFC curves with negative magnetization values (even though a positive measuring field is applied). The magnetic field of the magnetometer is measured by the voltage drop over two shunt resistors which account for different field ranges [12]. These circuits have to be carefully calibrated. A residual misfit of about 1% remains typically and can cause artifacts at about 350 mT. In addition to the field control the unipolar power supply can impair measurements in case of a zero current offset or trapped residual flux in the magnet itself.

A cascade of detection circuits is used, since the magnetometer has to measure magnetization values over a range of several orders of magnitude. Two factors are

---

[11] The respective curve shape was approximated by a polynomial.

[12] According to Quantum Design a direct measurement of the magnetic field is not installed due to the high field range which can not be covered by a single sensor.

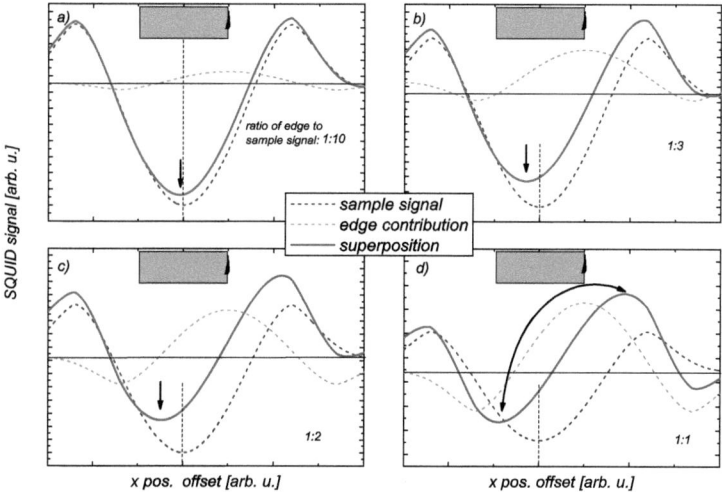

Figure 2.5: Shift of the x-position. Two magnetization contributions are superimposed. The blue curve simulates a diamagnetic response whereas the green curve mimics a ferromagnetic contamination at the edge of the sample. a) High magnetic field: Hardly any influence of the contamination. b) and c) with decreasing field values the diamagnetism weakens and the peak of the superposition is moved of lower x-positions. d) low fields: The ferromagnetic contribution starts to be the relevant magnetization peak. A jump to higher x-position will occur. Note that the sample does not extend to this position.

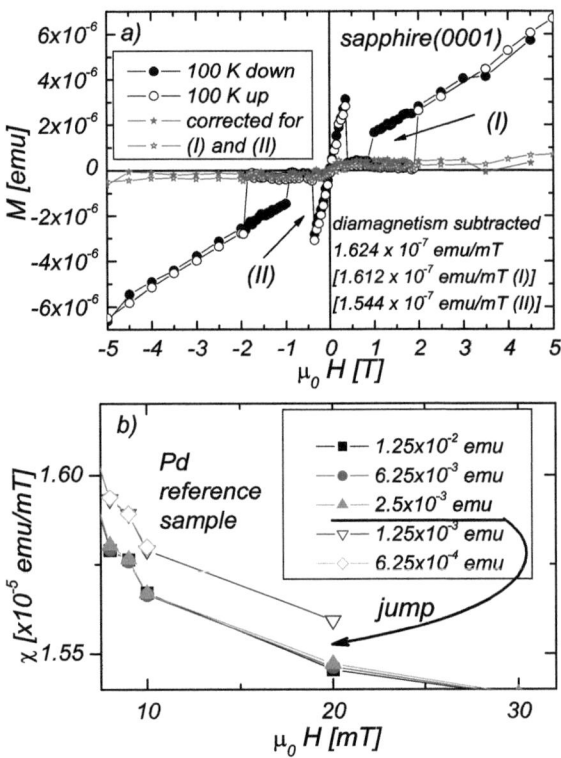

Figure 2.6: Influence of emu-range change. a) Hysteresis curve of pure sapphire revealing SQUID artifacts. Both types of artifacts can be removed by an adjusted diamagnetic background. b) susceptibility measurements of a Pd reference sample. Depending on the emu range, different magnetization values are measured.

adjusted: The *gain* changes the gain of the voltmeter card. The *range* changes the integration constant of the RF amplifier. During a measurement the software automatically chooses the proper settings. A hysteresis is normally taken over the full available field range, therefore respective adjustments of *gain* and *range* take place several times. While the *gain* is changed each step the *range* is change only every three steps. The change of the *range* is typically accompanied by a small step in the measured magnetization values [13]. After subtraction of a diamagnetic background the respective steps become obvious.

Figure 2.6 a) shows a hysteresis measured on pure sapphire. The two hysteresis like signatures appearing at field values between ±1-2 T (I) originate from the automatic adjustment of the *range*. The software tries to avoid a change of the *range*, therefore the hysteresis is purely caused by the software itself and conclusively the steps occur at different field values depending on whether the field is ramped up or down.

For small magnetic fields figure 2.6 a) shows an effect (II) caused by an above mentioned bad calibration of the magnetic field control. The step precisely occurs at the change from hi-res disabled to hi-res enabled mode of 350 mT.

Both artifacts in figure 2.6 a) can be removed by subtraction of an altered diamagnetic background as shown. Note that the respective percentages of change in the measured diamagnetism were reproducible on other samples. They are somehow intrinsic of the given SQUID magnetometer.

In figure 2.6 b) the magnetization change due to the different *range* settings is demonstrated with a Pd calibration sample. The respective detection sensitivities are given in the legend.

Finally it should be added that even after careful consideration of artifacts residual hysteresis like signatures might occur. During this work a detection limit of $4 \cdot 10^{-7}$ emu was assumed as recommended in [87].

2.4 XAS measurements at beamline ID12

Synchrotron measurements within this work were exclusively conducted at the beamline ID12 of the European Synchrotron Radiation Facility (ESRF) in Grenoble,

---

[13] Which is probably due to the change of the hardware, namely the R-C circuit responsible for integration.

France. This beamline is dedicated to polarization-dependent X-ray absorption and excitation spectroscopies in the energy range from 2.0 keV to 20 keV. The incident light is monochromatized by a Si(111) double crystal which is thermally stabilized at a temperature of 140 K. Some basic data of the undulators used by ID12 are given in table 2.4.

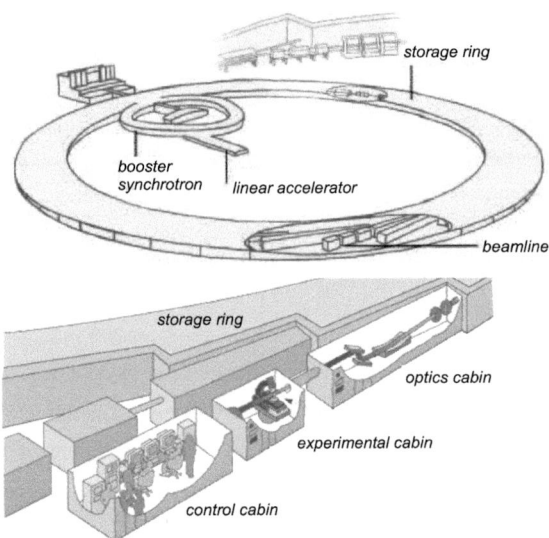

Figure 2.7: Schematic sketch of the ESRF and a beamline. Electrons are accelerated in a two step process: After energy gain in a linear accelerator the electrons are further powered in a cyclotron reaching their final energy. Afterwards injection into the storage ring with its tangential arranged beamlines takes place. Pictures are used by courtesy of the ESRF.

X-ray absorption spectroscopy - experimental details

All XAS spectra were measured by total fluorescence yield which enables bulk sensitivity contrary to electron yield measurements. The spectra were taken with respect to a reference intensity I_0 which was measured by metal meshes in front of the ex-

Undulator	APPLE-II	HEKIOS-II	EMPHU
Magnetic Period	38 mm	52 mm	80 mm
Number of periods	42	31	19
Photon Energy (1st harmonic)	5 - 9.1 keV	3.16 - 6.2 keV	1.6 - 4.35 keV
Brilliance	6.2×10^{19}	2.6×10^{19}	1.0×10^{19}
Helicity reversal time	< 5 sec	< 5 sec	160 msec

Table 2.1: Parameters of helical undulators of beamline ID12 at the ESRF [88].

perimental partition of the beamline. Depending on the absorption edge additional filters were installed in front of the photo diodes to reduce the background signal. XAS spectra have been measured with different experimental setups, depending on whether structure or magnetism of the samples should be investigated. This mainly affects the final part of the beamline, in particular the detector configuration and sample mounting.

All spectra have been normalized to the edge jump to yield comparability.

Absorption edge	energy[eV]	I_0	Filter
Zn K-edge	9659	Ti	-
Co K-edge	7709	Ti	Fe
Ga K-edge	10367	Cu	-
Gd L_3-edge	7243	Ti	Fe

Table 2.2: I_0 settings and detector filters depending on absorption edge. Electron binding energies are given according to [89].

X-ray linear dichroism - experimental details

Structural investigations by XLD have been conducted in a cube like vacuum system with 8 photo diodes placed in front of the sample roughly covering a hemisphere. The samples are fixed on a multiple sample holder which had a 360° rotatable axis perpendicular to the synchrotron light. Up to 7 samples can be glued on each side of the aluminum bar. By rotating the sample holder by 180° either the samples on the top or down side are measured. Typically an angle of 10° with respect to the synchrotron light was adjusted. This *grazing incidence* configuration is considered to be a good compromise between a maximized coverage of the sample by the beam and still fulfilling the geometrical needs in terms of crystal orientation [14]. A lateral motion of the sample holder enabled the positioning of the different samples roughly in the synchrotron beam. The final adjustment is accomplished by remote control of step motors moving the whole endstation.

The linear polarization of the light was achieved by a quarter wave plate consisting of a 0.9 mm diamond (111) plate. It transforms the circular polarized light generated by the undulator and monochromatized by the Si-crystal into linear polarized light by use of a forward Bragg reflection. Details can be found in [90].

Prior to the measurement a Kapton foil is inserted into the beam path. The foil is oriented such that a fully polarized fraction (Brewster angle) of the synchrotron light is reflected to a photo diode. The maximum and minimum of this signal is used to calibrate the quarter wave plate. The polarization by the quarter wave plate changes with the wavelength/energy of the incident light. Therefore the calibration of the angle of the quarter wave plate was preformed at the start and the end of the energy range of a respective energy scan. The angles of intermediate energy points have to be interpolated from these calibration points. During a XLD measurement the quarter wave plate is flipped at each energy point, minimizing possible artifacts of energy shifts.

XANES and XLD Simulation

Thanks to much enhanced computing power of ordinary computers, XAS-spectra can nowadays be simulated within a few days. For this work spectra were calculated with the FDMNES-code by Y. Yoly [91]. The calculation uses a multiple scattering formalism within a muffin tin approximation. Each atom is described by spatially separated spherical potentials. Empty interstitial regions are approximated by assuming a constant potential [92]. Due to potential backscattering events of the photoelectrons the probability of finding all scattering events within a region close to the photon absorbing atom is high. Therefore simulations can achieve very good agreement with experimental spectra already by limiting to a small cluster of atoms. Details can be found elsewhere [93]. XANES and respective XLD spectra for Gd:GaN and Co:ZnO were simulated. For the simulations of GaN the bulk values of the lattice constants $a = 3.189$ Å, $c = 5.185$ Å and $u = 0.377$ Å were taken together with a core hole lifetime of 1.82 eV for Ga [45]. In case of ZnO $a = 3.2459$ Å ,

---

[14] The inevitable occurrence of a residual XLD can be estimated to be about 1.5%

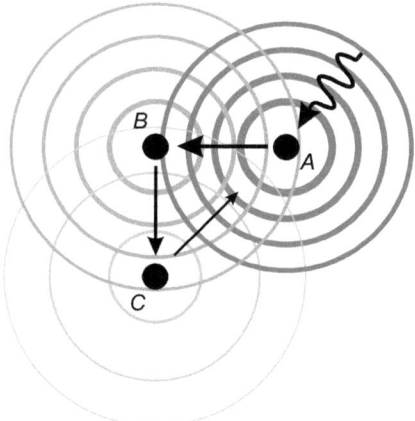

Figure 2.8: Schematical illustration of multiple scattering. Depending on its energy the photoelectron or the describing wave function, respectively can experience several scattering events in the close environment of the photon absorbing atom.

$c = 5.2069$ Å and a dimensionless u parameter of 0.382 was used [94]. The core hole lifetime was set to 1.67 eV according to [95]. Typical simulation cluster contain for example a supercell of $(Zn_{23}Co)O_{24}$ atoms. The calculation sphere is chosen up to 8.5 Å.

X-ray magnetic circular dichroism - experimental details

Circular polarized light is provided by the helical undulators of the beamline. Two different endstations were used to perform XMCD measurements. Predominantly the spectra were taken in a superconducting magnet setup which can provide magnetic fields up to 6 T. Two samples were typically installed on a copper wedge which fixed a 15° grazing incidence geometry. A liquid helium flow system allows to cool the sample holder down to 6 K. Normal incidence measurements were less frequently conducted with an alternative sample holder. A ring diode installed concentrically to the beam was used to measure the photon emission.

A second endstation was equipped with an electro magnet providing up to 0.6 T. This station was preferred for fast XMCD hysteresis measurements with high field resolution, since the magnet was driven by a bipolar power supply.

Chapter 3

Paramagnetic impurities in SiC, ZnO and Al$_2$O$_3$

The discussion in literature about DMS includes very dilute semiconductors with dopant concentrations below 10^{-16}/cm^3 [96]. The concentration of magnetically active impurities in grown layers and substrates are frequently of the same order of magnitude and thus have to be carefully considered with respect to magnetic properties of the material. Impurities in the host materials like ZnO or GaN can be interpret as the ultimate limit of dilution in terms of DMS. They are expected to be perfectly isolated inside the crystal matrix, which means there is no interaction of one impurity atom/ion with another. It has to be added that this discussion also includes all kinds of paramagnetic defects, i.e., vacancies, antisites or dislocations which result in a permanent magnetic moment. Respective ESR spectra are numerously reported in literature.

Figure 3.1 shows the temperature dependence of the magnetization of a commercial ZnO substrate measured at a field of 10 mT. Even though pure ZnO is not expected to show any paramagnetic signal, the magnetization increases at very low temperatures. This behavior results from the 1/T magnetization dependency (Curie law) of paramagnetic impurities. The Boltzmann-Energy at $T = 2$ K can be easily calculated, which gives 1.6×10^{-4} eV. Assuming a paramagnetic $S = \frac{1}{2}$-state the Zeeman-splitting will be of the order of $\approx 10^{-6}$ eV. Even though still smaller than the thermal energy the Boltzmann statistics results in:

$$\frac{N_a}{N_b} = e^{-\frac{g\mu_B H}{k_B T}} \approx 0.99 \qquad (3.1)$$

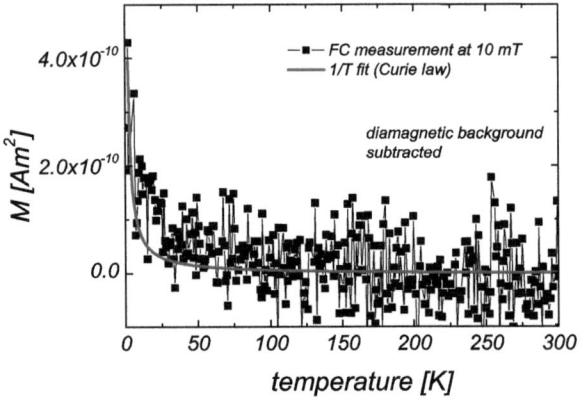

Figure 3.1: Increase of magnetization as measured by a SQUID at low temperatures due to paramagnetic impurities in a ZnO sample.

Element	Abundance
Si (Silicon)	10 ppm
S (Sulfur)	4 ppm
Fe (Iron)	2 ppm
Na (Sodium)	1 ppm
K (Potassium)	1 ppm
Ca (Calcium)	1 ppm
Ti (Titanium)	< 1 ppm
Y (Yttrium)	< 1 ppm
Zr (Zirconium)	< 1 ppm
Cr (Chromium)	< 1 ppm
total:	< 23 ppm

Table 3.1: Concentration of residual impurities in sapphire as provided by the Crystec company (ppm = part per million)[97].

with $N_{a/b}$ being the population number of the respective energy levels. It is this asymmetry of population which results in an overall magnetization even at small measuring fields of 10 mT. The general description of the magnetization due to population of Zeeman-split spin states is provided by the Brillouin function as introduced in section 1.1. Its temperature dependence yields the well known $1/T$ Curie-law.

In table 3.1 the residual contamination of sapphire substrates is tabulated. Note that most of the impurities are expected to be magnetically neutral, in other words without permanent magnetic moment per atom/ion. As will be shown in section 3.3 the Cr^{3+}-ion and the Mo^{3+}-ion are the most prominent contributions visible in magnetic resonance measurements.

3.1 Nitrogen in SiC

6H-SiC is the favored substrate to grow GaN or DMS based on GaN. The high conductivity of this material significantly aggravates ESR-investigations due to short circuiting of the electric field lines of the microwave reducing the quality factor and thus the sensitivity (see 2.2).

In figure 3.2 a typical spectrum of pure SiC is shown at T = 60 K in the field range that corresponds to $g = 2$. Three resonances can easily be distinguished. These three narrow lines can be attributed to nitrogen donors in the SiC host [98]. There are three different possible positions for the nitrogen atoms, two on cubic sites and one on a hexagonal site as can be seen from the ABCACB structure of 6H-SiC. The cubic positions lead to an enhanced hyperfine(hf) splitting ($I_{Nitrogen} = 1$) whereas the hf-splitting of the hexagonal place cannot be resolved. The latter is superimposed on the central line which leads to the enhanced intensity. The resonance field of the central line corresponds to a g-factor of 2.007 ± 0.005. More accurate measurements at 142 GHz were undertaken by E. N. Kalabukhova et al. where a g-factor of 2,0048 has been reported [99].

The two lines marked by black arrows can be identified as exchange pairs of nitrogen donors on cubic sites [100]. A detailed description of the spectra of nitrogen donors in different SiC crystal structures can be found in [98] and references therein. More recent ESR results for intrinsic defects of SiC can be found in [101].

As shown in [102] the nitrogen-spectrum considerably hinders the detection of ad-

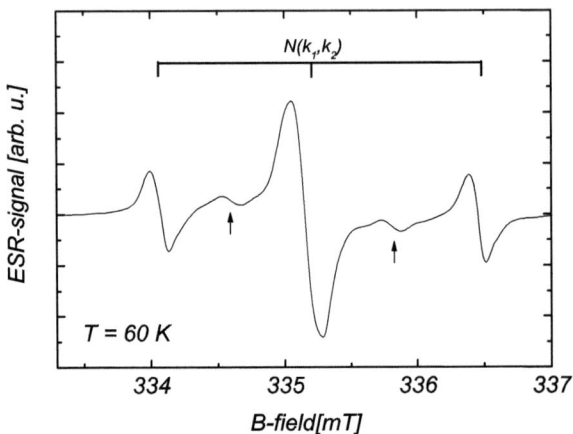

Figure 3.2: ESR-spectrum of nitrogen in a SiC substrate. The notation k_1 and k_2 denotes the two possible cubic places of the nitrogen donor in SiC. The resonance marked by arrows are cause by exchange pairs [100].

ditional signals close to $g = 2$, which is also within the field range where signals of paramagnetic Gd^{3+}-ions are expected to be found.

3.2 Transition metall ions in ZnO

In figure 3.3 a spectrum of a commercially available ZnO substrate [1] taken at $T = 300$ K is shown. The two major contributions stem from Fe^{3+} (red bars) and Mn^{2+} (blue bar) impurities. Both ions have a fivefold fine structure due to $S = \frac{5}{2}$. In case of Mn^{2+} the spectrum is further split due to the nuclear spin of $I_{Mn} = \frac{5}{2}$. The natural abundance of this isotope is 100%, therefore each fine structure line forms a sextet of equally spaced signals. In principle a fraction of 2.1% of the Fe line is also hyperfine split but not resolved due to the weak hyperfine splitting of the respective isotope (Fe^{57} with 2.1% natural abundance and $I = \frac{1}{2}$ [2]). Respective

---
[1] CrysTec GmbH, Kopenicker Str. 325, D-12555 Berlin, www.crystec.de
[2] $I = 0$ for all other relevant isotopes

TRANSITION METAL IONS IN ZnO

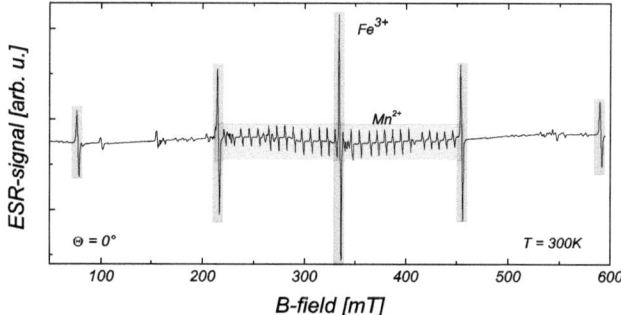

Figure 3.3: Fe^{3+}(red bars) and Mn^{2+}(blue bar) impurities in ZnO measured by ESR at $\nu = 9.3$ GHz. The magnetic field is oriented parallel to the crystal c-axis (0001). Both ions have a fivefold fine structure due to $S = \frac{5}{2}$. Contrary to the Fe lines the Mn resonances are sixfold hyperfine split by the nuclear moment $I_{Mn} = \frac{5}{2}$ leading to a spectrum of in total 30 lines. A detailed description of the spectrum has been given in [103, 104].

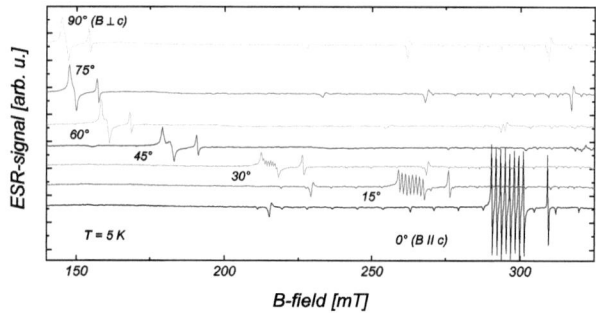

Figure 3.4: Co^{2+}-spectra in ZnO taken at T = 5 K for various angles. A group of 8 lines is clearly resolved for $\theta = 0°$. The line accompanying the group of Co^{2+} lines is assigned to Ni^{3+} [105]. Additional groups of small sixfold features are visible which originate from Manganese impurities. The corresponding spectra are highly saturated due to the long spin lattice relaxation time at low temperatures of Mn in crystal environment (P_{MW}= 2 mW) [106].

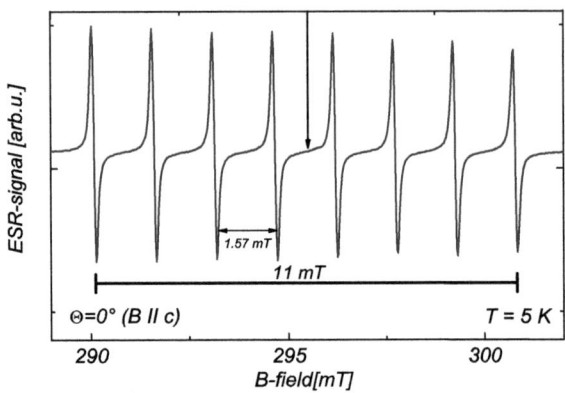

Figure 3.5: Co^{2+} ESR spectrum with well resolved hyperfine splitting at T = 5 K. The black arrow points to the center of gravity of the group of lines.

spectra have been reported in the literature [103, 104].

The first report of the ESR spectrum of Co^{2+}-ions embedded on cation sites in zinc oxide dates back to 1961 [107]. The ESR spectrum of this ion will be discussed in more detail, since Co-ions are supposed to activate ferromagnetism in ZnO.

Figure 3.4 shows scans taken at a temperature of 5 K in an out-of-plane geometry. This means that the external applied field angle was varied from $\vec{B} \parallel$ c-axis ($\theta = 0°$, out-of-(sample)-plane) to $\vec{B} \perp$ c-axis ($\theta = 90°$, in-(sample)-plane), since the sample is grown in c-axis(0001) direction.

The best spectrum for identification of the paramagnetic ion was taken at $\theta = 0°$. One can clearly see a group of eight equidistant resonance lines. This can be seen as an fingerprint of a nucleus with nuclear magnetic moment of $I_n = \frac{7}{2}$ causing a eight-fold hyperfine splitting($I_n(^{59}\text{Co}) = \frac{7}{2}$, 100% natural abundance). This spectrum is shown in greater detail in figure 3.5.

Assuming that the sample is rotated from the lattice axis c (0 0 1 (hkl), 0 0 0 1(hkil)) to the in-plane axis a (2 $\bar{1}$ 0 (hkl) (1 $\bar{1}$ $\bar{1}$ 0 (hkil)) one can simplify the Hamiltonian 1.28 to:

$$H_S = \mu_B \left(g_\| B \cos\theta\, S_z + g_\perp B \cos\theta\, S_x\right) + D\left[S_z^2 - \frac{1}{3}S(S+1)\right] \quad (3.2)$$

With $g_\| = g_{xx} = g_{yy}$ and $g_\perp = g_{zz}$

For the out-of-plane geometry the matrix of the Hamiltonian is diagonal and the energy of the eigen-states is:

$$E = D(m_s^2 - \frac{5}{4}) + g_\| \mu_B B m_s \quad (3.3)$$

Since one only observes energy-differences between states with $m_s = \pm\frac{1}{2}$ the resonance condition can be written as $\Delta E = \hbar\omega = g_\| \mu_B B$. Additionally, the hyperfine interaction has to be taken into account, which leads to the final expression:

$$\Delta E = \hbar\omega = g_\| \mu_B B + A_\| m_I \quad (3.4)$$

A more complicated situation arises for the in-plane geometry since a mixing of the lowest lying $E_\frac{1}{2}$ and $E_\frac{3}{2}$-states occurs. In this case one observes a resonance at the transition energy $\Delta E = \hbar\omega = g_{\perp eff}\mu_B B$. As shown in more detail in appendix A this can be approximated [108, 109] assuming $D >> \hbar\omega$:

$$g_\perp \approx \frac{1}{2}g_{\perp eff} \quad (3.5)$$

For this orientation the hyperfine splitting is strongly reduced leading to a superposition of the different m_I components. The respective splitting is described by $A_\perp$.

The green curve in Figure 3.6 shows the angular dependence of the hyperfine splitting. The black curve shows the respective resonance fields of the center of the group of lines. Both curves prove a uniaxial behavior. The red line presents a fit of the resonance field based on: $f(\theta) = P(g_{\perp eff}^2 sin(\theta)^2 + g_\|^2 cos(\theta)^2)^{-\frac{1}{2}}$ with P as fitting parameter. Obviously the angular dependency is well described by the g-tensor. Table 3.2 summarizes the derived values for the components of the g- and A-tensor.

With regard to DMS it has to be stressed that the observed spectra belong to purely isolated non-interacting ions. The following arguments are solely based on ESR observations.

First of all, the existence of a well resolved hyperfine splitting as shown in figure

	$g_\parallel$	$g_{\perp eff}$	$g_\perp$	$A_\parallel[10^{-4}\mathrm{cm}^{-1}]$	$A_\perp[10^{-4}\mathrm{cm}^{-1}]$
this work	2.247	4.557	2.278	16.4±0.1	3.3±0.2
literature [107]	2.2384	4.551	2.2768	15.9	2.9
	(2.243)		(2.2791)	(16.11)	(3.00)

Table 3.2: Results for g- and A-tensor for Co^{2+}. The error bar of the g-values is 0.005, which is mainly due to inaccuracy of the sample alignment. Literature values given in brackets are taken form the original work by T. Estle et al. [110].

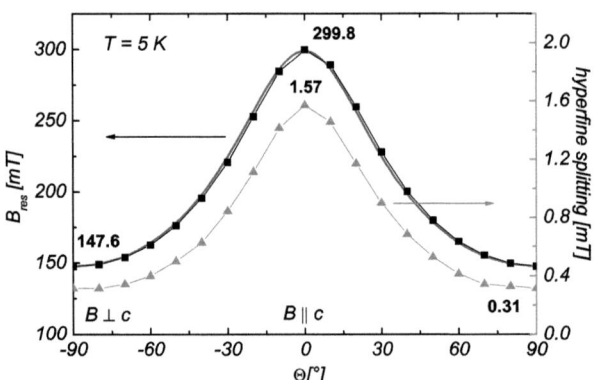

Figure 3.6: Angular dependence of B_{res} (Center of gravity) and hf-splitting. The red curve is a fit based on $f(\theta) = P(g^2_{\perp eff}sin(\theta)^2 + g^2_\parallel cos(\theta)^2)^{-\frac{1}{2}}$ with P as fitting parameter.

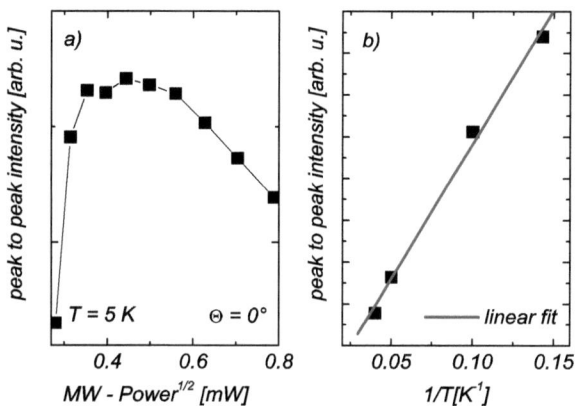

Figure 3.7: Paramagnetic microwave power and temperature dependence of the Co^{2+} ESR signal

3.5 would smear out in case of the presence of magnetic dipolar interactions arising from neighboring ions. This can be shown already for dopant fractions x > 0.001. Respective concentration dependencies for TM-ions in crystals have been reported in the literature [111].

A strong indication for the paramagnetic origin of the observed ESR resonance is shown in figure 3.7 a) and b). The absorption of microwave quanta means exciting electrons to an upper level. Therefore in case of high microwave powers a significant population of this level can be achieved [3]. The probability for a photon to be absorbed by an electron of the lower level or to induce an emission of a coherent photon from an electron of the higher level is equal [112]. This means in turn that for equally populated energy levels absorption and emission cancel out.

In figure 3.7 a) the square root [4] of the microwave power is plotted over the ESR-signal intensity. After a steep increase at low power the intensity reaches a maximum after which a further raise of the microwave power leads to a reduced absorption.

---

[3] It is obvious that the lifetime of excited states strongly depend on relaxation processes. Hence the population process is temperature dependent as is the saturation of the ESR signal.

[4] Only the magnetic field of the microwave accounts for absorption

The latter is a clear indication of the saturation of the energy levels. It is noteworthy that this effect hardly occurs in ferromagnetic resonance at the available MW power. Therefore figure 3.7 a) can also be used as a crosscheck for a ferromagnetic origin of the signal. Since the linewidth did not change in the observed temperature regime, the peak to peak amplitude of resonance curves is proportional to the intensity [113]. The latter is defined as the integral of the absorption curve which in turn is proportional to the magnetization. The temperature dependence of a Curie-like paramagnet is $M \propto \frac{C}{T}$. Therefore the linear behavior of the intensity values in figure 3.7 b) for T= 5, 10, 20 and 25 K also corroborates that Co^{2+}-ions in this sample are isolated, i.e. they behave as an ideal dilute paramagnet.

3.3 Cr^{3+} and Mo^{3+} in sapphire - Al_2O_3

Sapphire was typically used as a substrate for Co:ZnO growth. At room temperature all sapphire wafers show traces of paramagnetic impurities. The magnetic resonance spectrum of the most prominent one, Cr^{3+}, is shown in figure 3.8. The Cr^{3+}-ion as an impurity in Al_2O_3 is well know from optical and ESR studies.

High concentrations (%-range) of the ion cause the sapphire to appear reddish which is in this case known as ruby. Depending on concentration [114] and spectrometer settings the respective spectra could be very clearly observed at room temperature. Exemplarily three angles 0°, 60°, and 90° are plotted in figure 3.8. Additionally, the energy level scheme of this d^3-ion for the case $B \parallel c$ is shown. Note that the zero field splitting D is negative (D $= -2.37 \cdot 10^{-5}$ eV [52]), which leads to population of a $m_s=-3/2$ state at very low temperatures. According to the selection rule $\Delta\ m_s = 1$ no MW absorption is possible at very low temperatures. However, the fading of the lines at temperatures (see fig. 3.9) below 40 K has to have another origin, since $k_B T$ is still sufficient to populate the respective energy levels. For instance the change in relaxation times due to reduced phonon interactions can be considered [115, 116]. The latter has been studied in view of maser and laser applications of ruby [117]. For angles different from 0° the selection rules are not that strict any longer and forbidden transitions are visible due to a mixing of different states. A thorough treatise of the paramagnetic resonance spectra of Cr^{3+} can be found in [118]. Considering the concentration of Fe (2 ppm, see table 3.1) the absence of any Fe^{3+}

Cr^{3+} AND Mo^{3+} IN SAPPHIRE Al_2O_3

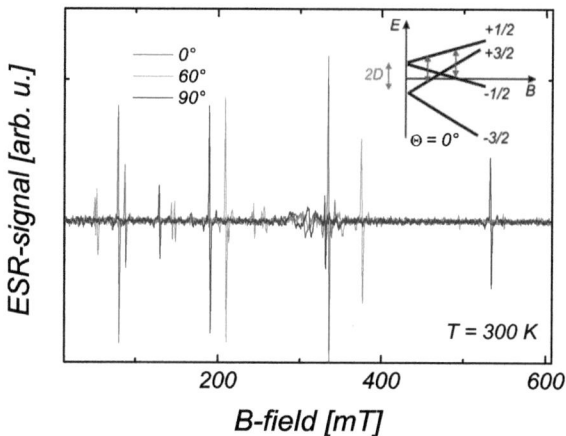

Figure 3.8: Paramagnetic resonances in sapphire at room temperature and ν=9.3 GHz. The most prominent signals originate from Cr^{3+}. An isotropic background was subtracted. The inset shows the splitting of the energy levels of Cr^{3+} for $B \parallel c$ ($\theta = 0°$).

signal at room temperature in our sapphire substrates is rather astonishing. Besides concentrations other than given in table 3.1, mixed valences as discussed e.g. in [119] can be considered as possible explanation.

Figure 3.9 shows a magnetic resonance signal at T = 10 K which could be found in many Co:ZnO films grown on sapphire at low temperatures. The angular dependence taken at 5 K within a range of 200° is shown in the inset and proves a uniaxial symmetry of the paramagnetic center. The blue line presents a fit based on: $f(\theta) = P(g^2_{\perp eff} sin(\theta)^2 + g^2_{\parallel} cos(\theta)^2)^{-\frac{1}{2}}$ with $g_{\perp\ eff} = 3.92$, $g_{\parallel} = 1.97$ and P the only parameter. Additionally to the central line in figure 3.9 a sixfold hyperfine structure is visible. This spectrum could be identified as substitutional Mo^{3+} in Al_2O_3 due to the intense central transition and the surrounding sextet of lines. The latter is due to the hyperfine interaction with ^{95}Mo and ^{97}Mo nuclei possessing total nuclear

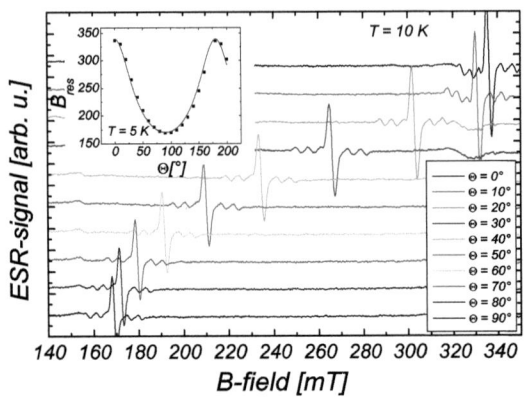

Figure 3.9: Mo^{3+} embedded in Al_2O_3. Note the absence of any Cr^{3+} signals at this temperature. The inset shows the uniaxial angular dependence of the resonance field for a 200° angular range.

spins of I = 5/2 resulting in almost similar splitting [5]. The natural abundances of these isotopes are 15.78% and 9.60%, respectively [6]. Therefore in total 25.38% of the Mo resonance spectrum will be sixfold. A double integration of the hyperfine and the central section of the spectrum yields the expected fraction of $\approx 1/4$. The ion is octahedrally coordinated on substitutional Al^{3+} site. It is isoelectric to the well studied Cr^{3+}-ion in sapphire. In figure 3.10 the temperature dependence of the spectrum is plotted. Apart from an expected 1/T decrease of the intensity for temperatures from 10 K on an increase from 5 K to 10 K is observed (see inset). Contrary to the case of Cr^{3+} discussed before, for Mo^{3+} in Al_2O_3 a reported zero field splitting of $D \approx -50$ GHz $= -2 \cdot 10^{-4}$ eV [120] might account for the decrease of the intensity. Due to the reduced thermal energy at T = 5 K ($\approx 2.3 \cdot 10^{-4}$ eV) which is of the same order of magnitude as D the s=1/2 levels will be significantly less populated (see equation 3.1 and figure inset of figure 3.8).

Two possible origins for the existence of the Mo^{3+} have to be considered. In refer-

---

[5] μ/μ_N=-0.9142 (^{95}Mo) and -0.9335 (^{97}Mo)
[6] Several isotopes of Mo without nuclear spin exists - ^{92}Mo, ^{94}Mo, ^{96}Mo, ^{98}Mo, ^{100}Mo

DISCUSSION: IMPURITIES

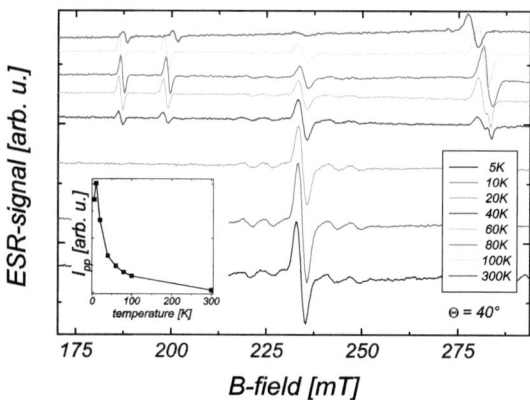

Figure 3.10: Temperature dependence of Mo^{3+} in Al_2O_3. The inset shows the peak to peak amplitude of the signal

ence [121] the Mo content of the sapphire was attributed to Mo containing crucibles which cause a contamination during the production process. Another source might be related to the growth of Co:ZnO samples, since the sample holders were made from Mo. For cleaning and annealing purposes the sapphire substrates were heated under UHV conditions above 1000°C prior to the growth process. Within this work no clear correlation could be established. The signal could also be traced in unheated sapphire substrates indicating that both possibilities can contribute. For temperatures of 40 K up to 300 K additional resonance signals due to Cr^{3+} develop (figure 3.10).

3.4 Discussion: Impurities

If the high sensitivity of magnetic resonance measurements is used to sense very small signals, one inevitably has to consider impurities of the substrate. Predominantly two types of substrate were used during this work: 6H-SiC and Al_2O_3. In case of Gd:GaN both substrates were used, whereas Co:ZnO was predominantly grown on sapphire. The ESR spectrum of SiC shows only one paramagnetic center present in the material which is interstitial nitrogen. This donor atom shows a characteristic signature at low temperature and high dilution in very good agreement with literature. As will be shown in chapter 4 several additional resonances occur after ion implantation in the same field range.

Sapphire substrates exhibited several paramagnetic resonances with changing spectral weight depending temperature. The most prominent contributions originate from Cr^{3+} and Mo^{3+}. Both ions are isoelectric and substitute for Al^{3+}. In hydrothermally grown ZnO substrates magnetic resonance signatures of Mn^{2+},Fe^{3+}, Ni^{3+}, and Co^{2+} were found. The concentrations of all these impurities were low, but the relatively high amount of material of the substrate wafers (compared to grown films of typically 100 nm thickness) enabled a detailed study of the Co^{2+}-ion as a paramagnetic impurity — the limit of dilution.

The main subject of this work was not the investigation of paramagnetic impurities. Dilute paramagnetic systems have been intensively studied in the middle of the last century. Theoretical descriptions date back to 1929 [53]. Nevertheless it turned out that a thorough understanding of these signals was indispensable for the search of eventual additional magnetic resonance signals in DMS. Even though the impurity signals generally compromise the spectra interpretation there are also beneficial aspects. First of all, the signals could be used as a reference for the proper angular alignment. Paramagnetic impurities provide a valuable cross check for the angular alignment, since the error-prone manual adjustment of the sample holder to the goniometer can cause significant angular offsets. Assuming a constant concentration and homogeneous distribution of the impurities in the substrates, resonance signals of impurities can also be used to estimate of the relative intensity of additional signals without using an additional spin reference inside the cavity. In the case of internal fields inside the sample paramagnetic impurities can even be used as field sensors by evaluating potential resonance field shifts [107].

Chapter 4

Experimental results for Gd:GaN

4.1 Preparation of Gd:GaN samples

The growth of all GaN and Gd:GaN samples used for this work was conducted by molecular beam epitaxy which has proven to be suitable to obtain GaN of high crystalline quality [45]. Different variants have been used, ammonia assisted or nitrogen plasma assisted growth. All Gd:GaN samples are c-plane (0001) oriented. The samples originate from four different institutions.

- GaN and Gd:GaN samples grown by ammonia assisted MBE on 6H-SiC (0001) were provided by the group of Prof. Ploog (Paul-Drude-Institut für Festkörperelektronik, Berlin). Conventional Ga/Gd effusion cells were used together with highly purified ammonia. For details see [65].

- N_2 plasma assisted MBE growth was used by the Group of Prof. A. Rizzi (IV. physikalisches Institut, Universität Göttingen). Sapphire (0001) serves as substrate in conjunction with a ≈ 2 μm thick GaN buffer layer grown by metal-organic chemical vapor deposition (MOCVD) [122].

- GaN films were grown by ammonia assisted MBE on sapphire (0001) at the Centre de Recherche sur l'Hétéro-Epitaxie et ses Applications of CNRS, France. Additionally a GaN buffer layer was Si doped to accommodate the lattice mismatch [123].

- For some of the samples the Gd was incorporated after MBE growth of GaN by focused ion beam (FIB) implantation. All implantations were conducted in the group of Prof. A. D. Wieck (Lehrstuhl für Angewandte Festkörperphysik, Ruhr-Universität Bochum). Details of the implantation process can be found e. g. in [124]

4.2 Paramagnetic signatures in Gd:GaN

This section starts with a set of Gd:GaN samples, which seem to be ferromagnetic as indicated by SQUID results. However further studies on these and additional samples do not corroborate this result but underline the paramagnetic character of Gd:GaN.

4.2.1 Gd:GaN grown on SiC

Figure 4.1 shows 2θ XRD-scans for GaN and Gd implanted GaN grown on 6H-SiC (0001). The characteristic reflections in the wide range scans can be identified with GaN and the substrate SiC. These scans also exclude the formation of secondary phases down to the sensitivity limit of this structural analysis method ($\approx$ 2-4 nm). The red curve shows the XRD scan of a GaN layer after focused ion beam implantation of $1 \cdot 10^{15}$ Gd^{3+}/cm^2 which corresponds to an average volume concentration in the layer of $1 \cdot 10^{20}$ Gd^{3+}/cm^3 and was the highest dose used for this series of samples. The latter is calculated by assuming a maximum penetration depth of the ions of 100 nm, which is estimated by implantation profile simulations using the SRIM code [81]. Details can be found in [125].

The inset of figure 4.1 shows the (0002) [1] and (0006) reflection of GaN and SiC on an enlarged scale, respectively. The black arrows point to additional features visible in the reflections of the implanted sample. The most intense additional peak (1) shows up with almost half of the intensity of the original GaN (0002) peak. Adjacent, closer to the GaN reflection a second additional peak (2) of reduced intensity is visible. The peaks appear at angles smaller than expected for the GaN (0002) reflection, indicative of an increased c-lattice constant of the implanted region of GaN or a coherent Gd:GaN phase, respectively [23].

These additional features close to the GaN(0002) reflection of the implanted sample

---

[1] In figure 4.1 the cubic Miller index system is used for better readability. In the text the appropriate hexagonal system (hkil) is used, where h, k and l are identical to the Miller index, and i is a redundant (i=-h-k) index.

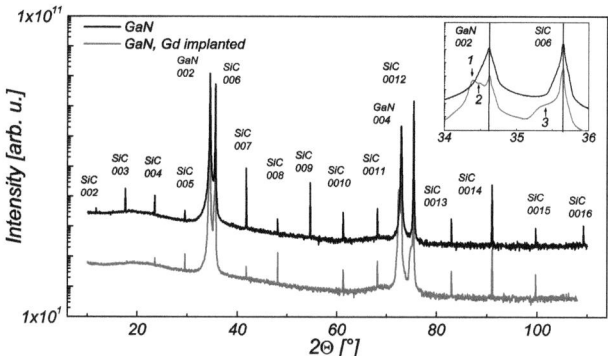

Figure 4.1: XRD-scan of hexagonal GaN and Gd implanted GaN. The lower scan was measured after implantation of $1 \cdot 10^{15}$ Gd^{3+}/cm^2 which corresponds to an a average volume concentration in the layer of $1 \cdot 10^{20}$ Gd^{3+}/cm^3. The inset shows structural implantation effects only visible for this highest dose.

are indicative of structural effects caused by the implantation process. Such effects are very likely, since in average every surface atom has been hit once at this high implantation dose. In literature a dose of $6 \cdot 10^{20}$ Ca^{3+}/cm^3 implanted in GaN with only 180 keV at liquid nitrogen temperature is reported to lead to amorphisation [126]. Arrow 3 in the inset points at an additional shoulder of the SiC (0006) reflection which is much lower in intensity than the prior discussed features. Nevertheless it suggests implantation effects in the substrate. Such an effect is hardly expected from SRIM simulations, since the layer thickness of this sample is about 600 nm. This shoulder may indicate a much deeper penetration of the implanted ions, which might be possible because the simulation software does not account for channeling effects or inhomogeneities, like screw dislocations of the GaN [96]. Another possible explanation is the formation of phase separated GdN which (002) reflection is around to 36.1° [127]. Possible crystal distortions due to strain might alter the lattice constant.

Figure 4.2 a) compiles the M(H) curves of two Gd implanted and one unimplanted GaN sample as they were measured with a SQUID at T = 2 K. Note that usually such hysteresis curves are taken as indicative of ferromagnetism. However, as will be shown in the following, a thorough analysis of results of complementary experimental methods has casted strong doubts on the origin of the signal, i.e. whether these M(H) curves reflect intrinsic ferromagnetism of Gd:GaN.

The most pronounced hysteresis (red diamonds) is found in the sample with 10^{20} Gd^{3+}/cm^3, which exhibits changes in the XRD-reflections in figure 4.1. The second Gd:GaN sample (green down triangles) was implanted with a much lower dose resulting in a layer concentration of $1 \cdot 10^{16}$ Gd^{3+}/cm^3. Both implanted samples show an open hysteresis which barely reaches full saturation at fields of 5 T (inset). The remanence of the hysteresis differ roughly by 1/3 and the saturation magnetization by less than a factor of two even though the implantation dose differs by four orders of magnitude. This surprising result was explained in literature within a sphere of influence model [65, 96] and will be referred to in more detail later on. Note that the slight slope at 2 K of the unimplanted reference sample is due to paramagnetic impurities of the SiC substrate

Besides the features suggestive for ferromagnetism of the implanted films like remanence and coercivity, the inset of figure 4.2 a) shows the strong curvature of the magnetization on a large field scale - far beyond field values where the two branches of the hysteresis are separated. At high fields the shape rather resembles a Brillouin or Langevin function and is therefore indicative of para- or superparamagnetic contributions.

The same samples were measured at room temperature revealing a much more step-like hysteresis (inset of figure 4.2 a) which corroborates the assumption of a superposition of ferromagnetic and (super)paramagnetic contributions at low temperatures. Results are compiled in table 4.1 and reported in [125].

This set of samples, was also investigated by ESR.The first and unexpected result to be mentioned is the absence of any ferromagnetic-like resonance at room temperature in these samples. All these film were grown on highly conductive SiC. The high conductivity of the samples lowers the quality factor of the setup which obviously hampers observations of small signals. Nevertheless, a ferromagnetic behavior of a 100 nm layer can typically be easily detected. Due to carrier freezing the sensitivity of magnetic resonance measurements is much enhanced at low temperatures. Angular-dependent measurements are expected to facilitate the identification of ferromagnetic resonance lines due to the expected crossing of the field value for $g = 2$

PARAMAGNETIC SIGNATURES IN Gd:GaN

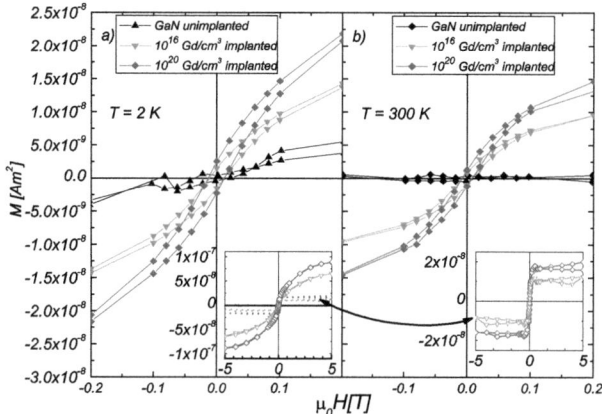

Figure 4.2: Hysteresis measurements on Gd implanted hexagonal GaN. The diamagnetic contribution of the substrate has been subtracted. a) Low temperature hysteresis of Gd implanted hexagonal Gd:GaN in comparison with unimplanted GaN. The inset shows an enhanced field scale showing the increase of the saturation magnetization and curvature with implantation dose. The dotted lines are T = 300 K data shown for comparison. b) Room temperature hysteresis of Gd implanted hexagonal Gd:GaN. The inset shows an enhanced field scale revealing the increase of the saturation magnetization with implantation dose.

("$\frac{\omega}{\gamma}$"), resulting from ferromagnetic anisotropies.

At low temperatures the ESR spectra were mostly governed by the nitrogen impurities of the SiC substrate as described in chapter 3. Measurements at T = 5 K for ion implanted Gd:GaN and Gd:GaN grown by MBE are shown in figure 4.3. The magnetic field range is limited to values corresponding to g = 2, since no additional signals were found in the range from 0 to 1.2 Tesla. The latter is consistent with the high purity of the samples. Paramagnetic Gd^{3+}-ions in a $^8S_{\frac{7}{2}}$ state are expected to show a signal close to g = 2 with small angular dependence due to the screening of the local 4f-electrons by the outer d- and s-electrons [52] and a seven fold fine structure can be expected from results of Gd^{3+} in ZnO [128].

The upper curve (black) of Figure 4.3 shows a typical spectrum of GaN grown on

Gd/cm^3	M_r (2 K)[Am^2]	H_c (2 K)[T]	M_{5T}(2 K)[Am^2]	M_{dia}[Am^2/T]
	M_r (300 K)[Am^2]	H_c (300 K)[T]	M_{5T}(300 K)[Am^2]	
$2.4 * 10^{16}$	$1.5 \cdot 10^{-9}$	$1.2 \cdot 10^{-3}$	$6.5 \cdot 10^{-8}$	$8.421 \cdot 10^{-7}$
	$0.9 \cdot 10^{-9}$	$0.8 \cdot 10^{-3}$	$1.1 \cdot 10^{-8}$	
$1 * 10^{20}$	$2.4 \cdot 10^{-9}$	$1.3 \cdot 10^{-3}$	$8.8 \cdot 10^{-8}$	$7.985 \cdot 10^{-7}$
	$1.2 \cdot 10^{-9}$	$0.8 \cdot 10^{-3}$	$1.8 \cdot 10^{-8}$	

Table 4.1: Results of magnetization measurements by SQUID on Gd implanted GaN.

SiC. Three resonances can easily be distinguished, whereas a much broader fourth one can be presumed because of the shifted ordinate values on the right and left of the central resonance, respectively. The spectrum stems from the 6H SiC substrate, which was already described in 3.1. The three narrow lines can be attributed to nitrogen donors in the SiC host. The broad superimposed resonance is likely to stem from residual free carriers, since the SiC substrate is highly conducting.

The red curve in Figure 4.3 shows the altered ESR spectrum for the sample implanted with $1 \cdot 10^{20}$ Gd/cm^3. The corresponding ESR-spectrum of the nitrogen interstitials is almost totally covered by a broad feature with a peak to peak width of approximately 0.8 mT. Superimposed on this dominating feature one can find three additional small peaks, labeled A, B, and C.

The same peaks but less pronounced are also visible in a sample that was implanted with 1/50 of the Gd amount (green curve). One can easily estimate that the ratio of the peak-intensities does not match the Gd concentration ratio which makes a correlation with the Gd content unlikely. The angular dependence of the nitrogen lines prevents further analysis of the small signatures A, B, and C, since they are not traceable in the full angular range.

The origin of these features might be implantation-induced defects. This is further corroborated by a spectrum of the annealed version of the sample with the highest implantation dose (light blue). All three peaks vanish after heating to 1300 K - most likely due to annealing effects improving the crystal structure, causing the paramagnetic centers to disappear. Indeed the spectrum of the annealed sample strongly resembles the pure GaN as does the spectrum of a additionally shown MBE grown sample with a concentration of $2 \cdot 10^{19}$ Gd/cm^3 (dark blue). The probability of developing a certain defect could be related to the actual crystal quality, i.e. the

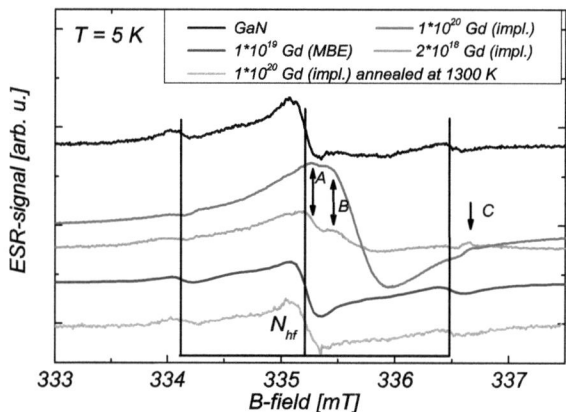

Figure 4.3: Comparison of the ESR spectra at g = 2 in out-of-plane geometry ($B \perp c$) of hexagonal GaN, highly Gd implanted GaN, and MBE-grown Gd:GaN. The three nitrogen hyperfine lines observable in the unimplanted GaN are hardly detectable in the highly implanted case due to a superimposed broad resonance. Note that scans were adjusted by the nitrogen spectra for better comparability. The ordinate is not to scale.

formation saturates during the implantation process. This might explain why these defects do not scale with the Gd dose.

In [129] no altered magnetization behavior of the annealed version of the highest implanted sample is reported. Therefore one can conclude that the additional paramagnetic centers as observed by ESR do not contribute to the ferromagnetic-like magnetization behavior, i.e. defect-induced ferromagnetism is unlikely.

4.2.2 Gd:GaN grown on sapphire (0001)

A second set of samples of GaN grown by ammonia assisted MBE on sapphire (0001) substrates instead of SiC was subject to Gd ion-implantation and identical investigations as mentioned before. The main intention of the examination of these samples was to confirm the previous results.

SQUID hysteresis measured at T = 5 K and T = 300 K are shown in 4.4 a). As discussed in section 2.3 the remanence of the hysteresis (inset) can be considered to be within the artifact level. Thus the ferromagnetic signature of the previous samples could not be reproduced. In figure 4.4 b) the XRD spectra of implanted and unimplanted GaN on sapphire are presented. As already seen for the samples grown on SiC the structural effects become evident at high doses. Note that the FWHM of the (002) reflection of both series of GaN layers is about 0.04°, which indicate a comparable high crystal quality. The additional shoulder visible in figure 4.4 of the sample grown on sapphire is less pronounced compared to the sample grown on SiC which might indicate less structural damage.

As discussed before contributions of the SiC substrate to ESR spectra considerably

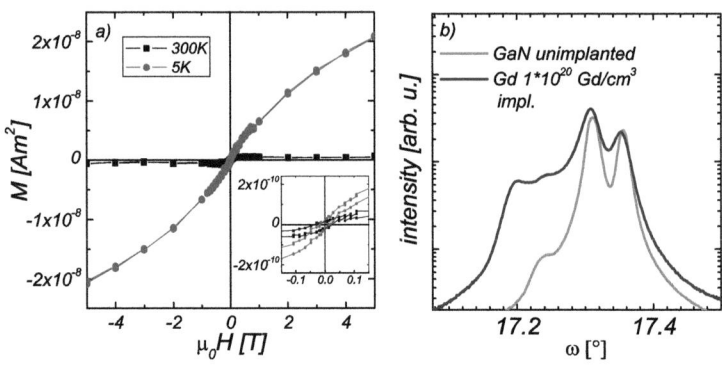

Figure 4.4: Structural and magnetic characterization of Gd implanted GaN on a sapphire substrate: a) SQUID-magnetometry results on Gd:GaN and implanted with $1 \cdot 10^{20}$ Gd/cm^3. The S-shape of the low temperature M(H) curve indicates paramagnetism. The inset shows a residual remanence and coercivity of the hysteresis which is likely to be an artifact as described in [87]. b) XRD spectra of GaN and Gd:GaN grown on sapphire. The double-peak structure originates from the X-ray source itself, namely Cu K$_{\alpha_1,\alpha_2}$. The additional reflections due to implantation are similar to samples grown on SiC (see figure 4.1).

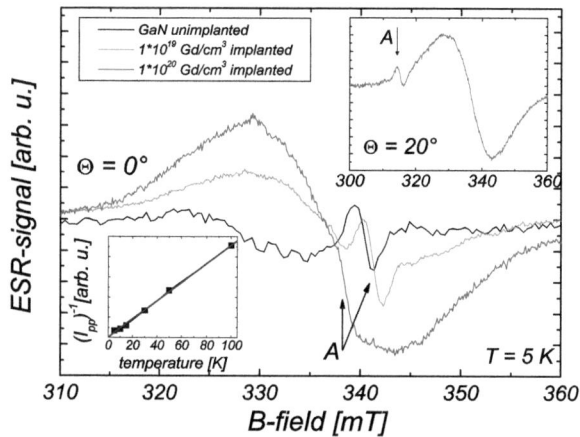

Figure 4.5: ESR spectra of Gd implanted GaN grown on sapphire at $\theta = 0$ and T = 5 K. A additional resonance with a peak to peak width of 15 mT is visible in the implanted samples. The signal is slightly distorted by the cavity background and an additional narrow resonance around 340 mT. The upper inset shows the shifted narrow line visible at $\theta = 20°$. In the lower inset the temperature dependence of the inverse of the peak to peak intensities of the broad resonance is plotted. The linear behavior is visualized by a fit (blue line). Small deviations at T = 5 K are due to saturation effects.

aggravate useful information gain. Therefore samples grown on a different substrate were likely to yield further insight. First of all the absence of a FMR like signal at room temperature was confirmed, which is in this case in agreement with the SQUID results.

The absence of the nitrogen lines in low temperature magnetic resonance spectra as shown in figure 4.5 leads to a very different spectrum compared to the signals of the samples grown on SiC (see figure 4.3). The resonance curve of the unimplanted GaN sample will be discussed first. A narrow line (black arrows) – in the following referred to as A – is visible and originates most likely from the buffer layer of Si doped GaN whereas the broader feature around 325 mT is known to stem from the cryostat. Angular-dependent measurements have only shown the narrow resonance to shift with angle as one may expect, since the cryostat is not rotated. This resonance is tentatively assigned to Si^{3+}(see e.g. [130, 131]), in the GaN buffer and can also be found in the implanted samples.

In the sample with the highest dopant concentration the resonance A and the cavity signal are masked in the $\theta = 0$ scan, because the spectrum is governed by an intense broad resonance centered at a field value of 337 mT which yield a g-factor of 2.00. The peak to peak width is approximately 15 mT which therefore is clearly distinct from less broad resonances discussed before in for Gd:GaN on SiC (figure 4.3). Angular-dependent measurements indicate an isotropic behavior within the accuracy of the measurement. The green curve in figure 4.5 shows the spectrum of a sample with a reduced implantation dose of $1 \cdot 10^{19}$ Gd/cm^3. As shown in the upper inset of figure 4.5 the signal A becomes also visible in the high implanted sample for altered angles. The signal A can be used as intensity reference, since all samples stem from the same wafer with the same buffer layer. The relative peak to peak intensity of the main broad resonance in both implanted samples turns out to be 0.16, which is of the order of the Gd implantation ratio of 1/10 (The slightly higher value is likely to stem from the cryostat contribution which naturally does not scale with sample size and acts as an offset in both samples.).

To clarify the origin of the magnetic resonance signal, temperature-dependent measurements were conducted. The lower inset of figure 4.5 shows the temperature dependence of the inverse of the peak to peak amplitude of the resonance signal. This amplitude is proportional to the intensity of the signal and thus to the magnetization, since the linewidth does not change noticeable with temperature. The values are well described by a linear fit crossing the point of origin. This $\chi \propto 1/T$ dependence corroborates pure paramagnetic behavior. Small deviations at T = 5 K stem from the onset of signal saturation.

The g-value, the absence of a strong angular dependence, and the temperature dependence of the peak to peak intensity of the above discussed signal could be interpreted in terms of a $4f^7$-ion in the $S_{7/2}$ ground state like Gd. In addition the signal scales roughly with the implantation dose.

However, this spectrum is different from what is reported in literature [132] and calculated from theory [52] for substitutional Gd^{3+} in GaN. The $4f^7$-ion in the $S_{7/2}$ ground state is reported to show seven well separated fine structure lines, observable

PARAMAGNETIC SIGNATURES IN Gd:GaN

at X-band frequency and described generally by:

$$H = g\mu_B HS + B_2^0 O_2^0 + B_4^0 O_4^0 + B_4^3 O_4^3 + B_6^0 O_6^0 + B_6^3 O_6^3 + B_6^6 O_6^6 \quad (4.1)$$

Here the influence of the crystal field is described by parameters B_n^m and the Steven's spin operators (see e.g. [52], table 16). The absence of such an energy splitting and respective spectrum indicate random, non substitutional incorporation of Gd (possibly due to the implantation process), which makes respective energy level calculations obsolete [2]. Moreover extensive line broadening of Gd due to interaction with implantation defects should be taken into account. As an alternative source of the resonance shown in figure 4.5 one may consider paramagnetic defects or conduction band resonances, both could be induced by the implantation. A conduction band resonance is rather unlikely as discussed in [133], since the resonance shows no temperature-dependent broadening. A potential paramagnetic defect in turn should also contribute to the magnetization as it is measured by SQUID (figure 4.4). If one calculates the maximum magnetic moment originating form the amount of Gd in the sample:

$$M_{Gd}[Am^2] = 1.6 \cdot 10^{14} \times \frac{7.93\mu_B}{7.93 \times 9.274 \cdot 10^{-24} Am^2} = 1.2 \cdot 10^{-8}\ Am^2 \quad (4.2)$$

The magnetization measured at T = 5 K will reduce to $\approx$ 90% due to thermal excitations, which can be derived from the Brillouin function. The magnetic moment as measured by SQUID is about $2 \cdot 10^{-8}$ Am². Therefore the maximum Gd contribution can only account for roughly the half of this. Thus a significant magnetization contribution of paramagnetic defects caused by the implantation is likely.

4.2.3 Element specific investigations of Gd:GaN

As can be seen from the results on the different samples of Gd:GaN the role which Gd plays for the magnetism in this dilute magnetic semiconductor is not understandable in a straightforward way. Therefore element specific measurements were carried out to clarify magnetic as well as structural properties.

XMCD and XLD measurements were conducted at the L_3-edge of Gd at the

---

[2]Similar resonance signals were found for Gd:ZnO grown by reactive magnetron sputtering (RMS).

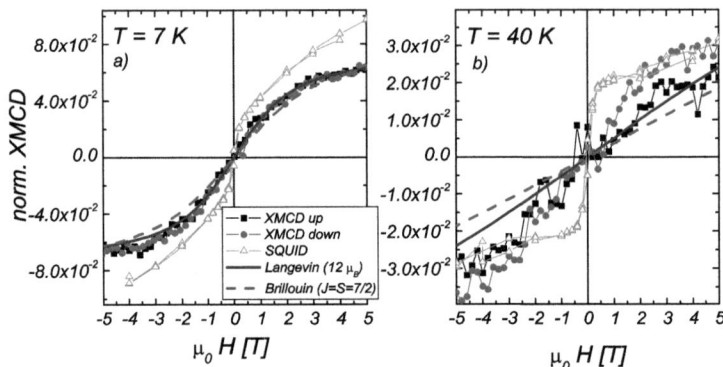

Figure 4.6: Element specific hysteresis taken at the Gd L_3-edge of a sample with nominally $2 \cdot 10^{19}$ Gd/cm^3. The terms *up* and *down* refer to the sweep direction of the magnetic field. Fits with Brillouin or Langevin functions at T = 7 K fail at T = 40 K. The saturation of the additionally plotted SQUID hysteresis is adjusted to the T = 40 K data. Assuming the same scaling factor as for the XMCD hysteresis to account for the T = 7 K data leads to a mismatch, which originates from additional magnetic contributions.

ESRF at beamline ID12. The Gd absorption edge of FIB implanted samples with the highest doses turned out to be barely detectable. A Gd implantation dose of $1 \cdot 10^{15}$ Gd/cm^2 (= $1 \cdot 10^{20}$ Gd/cm^3) on an area of typically 4 x 4 mm^2 is equivalent to solely $1.6 \cdot 10^{14}$ atoms (distributed in layer of $\approx$ 100 nm thickness). This is at the detection limit of synchrotron measurements which can be sensitive down to the mono layer range depending on the background signal.

Synchrotron measurements were preferably conducted on samples grown by MBE, since room temperature ferromagnetism as measured by SQUID was also reported for these [96].

In figure 4.6 a) and b) the field dependence of the XMCD (red and black symbols) at the Gd L_3-edge at two different temperatures for an MBE-grown, nominally $2 \cdot 10^{19}$ Gd/cm^3 doped sample is shown. The scans were taken at grazing incidence

PARAMAGNETIC SIGNATURES IN Gd:GaN

(15°) by total fluorescence yield as described in more detail in chapter 2.4 and published in [134]. The magnetization curves in a) resemble paramagnetic behavior. The pink dashed and blue solid curves are Brillouin and Langevin fits, respectively. The Langevin fit using a magnetic moment of 12 μ_B seems to represent the functional dependency slightly better but both fits fail to account for the T = 40 K data presented in the right plot. Therefore Gd can be excluded to behave purely paramagnetic in this sample. This discrepancy of the temperature dependence will be discussed in more detail in section 4.4.

In figure 4.6 additional SQUID measurements of the respective sample are plotted (green triangles). The data were normalized to the saturation value of the XMCD measurement at T = 40 K. The difference in the shapes of the M(H) curves between XMCD and SQUID indicates additional magnetization constituents. The 7 K SQUID data are normalized with respect to the 40 K XMCD- data. Therefore, the difference of the magnetization between SQUID and XMCD data in a) demonstrates magnetic contributions apart from Gd. This result indicates either significant influences of the host crystal, the GaN matrix, or a magnetic contamination of the sample.

This becomes even more obvious from the data in figure 4.7 which were taken with a magnet setup with higher field resolution but lower maximum field than in figure 4.6. Figure 4.7 a) shows the XANES and XMCD at the Gd L_3 absorption edge of the same sample at T = 295 K (red curve). The dopant atom apparently does not contribute to the room temperature ferromagnetic-like signal which is derived from SQUID-measurements, since there is no dichroism visible at this temperature. The lack of dichroism corroborates pure paramagnetic behavior of the dopant above the Curie-temperature of bulk like Gd ($\approx$ 293 K). This yields further doubt on a direct influence of Gd on the high temperature magnetization.

The green curve in figure 4.7 a) shows the XMCD at a temperature of T = 15 K. This temperature was chosen for a XMCD hysteresis measurement, which is depicted in b). Because of the high field resolution of the magnet setup the experiment can account even for small coercive fields. Within the noise level of the data a remanent magnetization and a coercive field stemming from the Gd-atoms can be excluded.

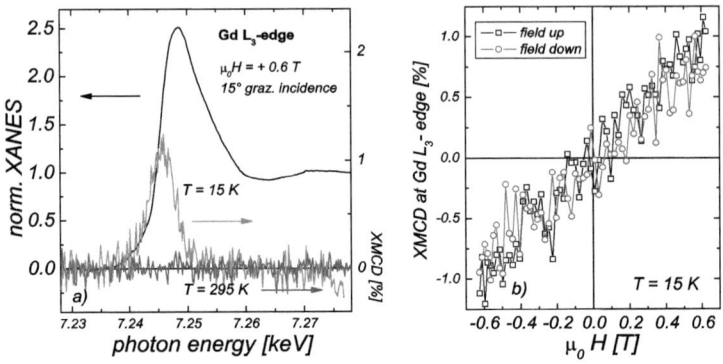

Figure 4.7: a) Gd L_3 absorption edge for grazing incidence and XMCD at T = 15 K and T = 295 K. No dichroism is visible at room temperature. b) XMCD hysteresis taken at T = 15 K with a high field resolution showing no remanence of the dopant. Within the accuracy of the experiment neither a remanence nor a coercivity is detectable.

4.3 Matrix polarization and magnetic polarons

So far, room temperature ferromagnetism of hexagonal Gd:GaN is predominantly claimed based on integral magnetization measurements by SQUID magnetometry [23]. The calculation of the magnetization contribution per Gd atom for the lowest implantation doses yields an astonishing number of up to 8000 μ_B [125]. This in turn means that Gd with an atomic moment of 8 μ_B itself can just account for $1/1000^{th}$ of M. The absence of ferromagnetic signatures in magnetic resonance measurements and the paramagnetic behavior of the dopant atom itself, further stresses the importance of the GaN host. A coalescence model suggested by Dhar et al. [65] assumes a polarization of the GaN matrix around the Gd atoms. According to this model long range magnetic order in the material depends on the overlap of the respective polarization spheres. From the lowest dopant concentrations reported (10^{16} Gd/cm^3) [65] resulting in a ferromagnetic behavior, this sphere of influence can be calculated to have an enormous size of 28 nm. For Ga atoms inside this sphere this would mean a average polarization of $1.1 \cdot 10^{-3}$ μ_B per atom. To clarify this, element specific XMCD measurements at the Ga K-edge and Ga L-edge have been performed.

An MBE grown sample with a nominal concentration of $2 \cdot 10^{19}$ Gd/cm^3 was chosen, which suggests room temperature long range magnetic order when measured by SQUID. Note that at this concentration practically all Ga atoms are assumed to be polarized, i. e. within a sphere of influence according to the model by Dhar et al.. The major challenge of XMCD measurements was to rule out artifacts, since the claimed polarization of Ga is very small. As described in section 2.4 helicity and magnetic field direction were both reversed, to account for this. The reliable detection of a possible Ga polarization requires great care.

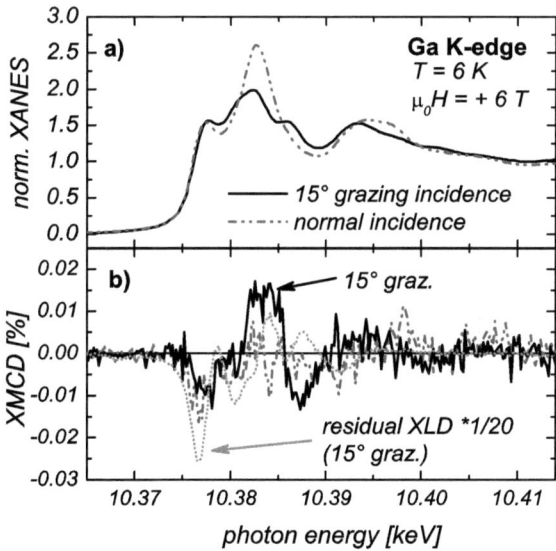

Figure 4.8: XAS and XMCD at the Ga K-edge. a) Absorption spectra for grazing and normal incident b) Very small XMCD signal after correction of 20 times larger residual XLD artifacts for grazing incidence. Qualitative similar spectral features for normal and grazing incidence support a very small magnetic polarization of Ga.

It should be emphasized that the circular polarization transfer rate of the X-ray beam which is monochromatized by a Si(111) double crystal monochromator is about 98%. A small linear polarization component at 45° (P2 Stokes - Poincaré parameter) is also generated by the monochromator at this energy and is of the order of 3%. By reversing the phase of the helical undulator, not only the circular polarization of the monochromatic X-ray beam is altered (helicity is reversed) but also there is a significant change of P2 [135]. The c axis of the sample is never aligned perfectly parallel to the vertical component of the light when looking into the light source, thus this situation leads to a contamination of the experimental spectra with a linear dichroic signal. However, this residual XLD signal does not reverse with the magnetic field, and thus, can be analyzed separately.

The residual linear dichroic effects are suppressed best with the c-axis oriented parallel to the incidence beam, i.e. normal to the sample surface of a c-plane wurtzite crystal.

The different fine structure of the absorption edge for grazing and normal incidence can be seen clearly in figure 4.8 a). In b) the XMCD spectra for both orientations are shown together with the residual XLD of the grazing incidence measurement, which was derived from spectra of opposite magnetic field directions. Note that the XLD amplitude is 20 times larger than the XMCD, thus indicating that the signal size is close to the detection limit. Only the similar spectral shape for grazing and normal incident makes the XMCD credible. A very small dichroic signal of 0.013% can be derived from both measurements.

The expected size of the XMCD signal at the K-edge, which is a measure of the orbital fraction of the magnetic moment at the Ga 4p states, can be estimated. From SQUID measurements, an overall magnetic polarization inside the spheres of influence of $1.1 \cdot 10^{-3}$ μ_B can be derived. The XMCD results can be compared to recent experimental data recorded at the Ga K-edge of InGaMnAs [136]. Here, a Ga 4p orbital moment of $8(4) \cdot 10^{-5}$ μ_B has been correlated with a maximum XMCD intensity of 0.05%.

Using this number for comparison with the XMCD in figure 4.8 b) the polarization can be roughly estimated to be at most $1 \cdot 10^{-5}$ μ_B. Note that in reference [136] the XMCD consists of only one positive feature resulting in a finite integral. In contrast in figure 4.8 b) the integral will be close to zero due to positive and negative contributions. Therefore the inferred magnetization is an *upper* bound for the Ga polarization.

This polarization is at least one order of magnitude smaller than what is inferred from SQUID measurements. The size of the XMCD signal at the Ga K-edge is therefore too small to corroborate the sphere of influence model.

Additional XMCD measurements at the Ga L_3-edge have further corroborated the absence of any polarization of the order of 10^{-3} μ_B [137].

4.4 Phase separation and clustering

In this section attempts are made to find extrinsic origins of ferromagnetic signatures of Gd:GaN. The first part presents experimental indications of phase separation in samples which are potentially free of secondary phases as found by XRD. The second part deals with a sample with evidenced clustering.

4.4.1 Indications of phase separation in Gd:GaN

Figure 4.9 a) shows the magnetic resonance scans at T = 5 K as a function of the polar angle for a MBE-grown sample with a Gd-concentration of $2 \cdot 10^{19}$ Gd /cm^3, which was already discussed in figures 4.6, 4.7, and 4.8. Around magnetic field values of 335 mT the sample holder and nitrogen donors in the SiC substrate cause a strong signal which saturates the detector diode due the chosen spectrometer settings. This part of the spectrum was already discussed in sections 3.1 and 4.2.

Within the remaining range of the spectra several magnetic resonance signals can be observed. The most prominent appears at 240 mT for an in-plane geometry, i.e. B⊥c and is referred to as Res 1 in the following. This signal was also found in a $1 \cdot 10^{18}$ Gd /cm^3 doped sample but with strongly reduced intensity. Plot b) of Figure 4.9 shows the angular dependence of the resonance field in the range where the signal was traceable revealing uniaxial behavior. Corresponding g-values change from 1.9 to 2.8, which suggests a significant influence of the crystal field if a paramagnetic origin is assumed. In addition the angular dependence of a second resonance field (Res 2) is partly shown. Note that the two signals have an opposite anisotropy, making a paramagnetic origin unlikely. Res 1 was subject to more detailed investigations. Figure 4.9 c) shows resonance field (red squares), full width at half maximum (FWHM)(green triangles), and intensity (black circles) of the line evolving with tem-

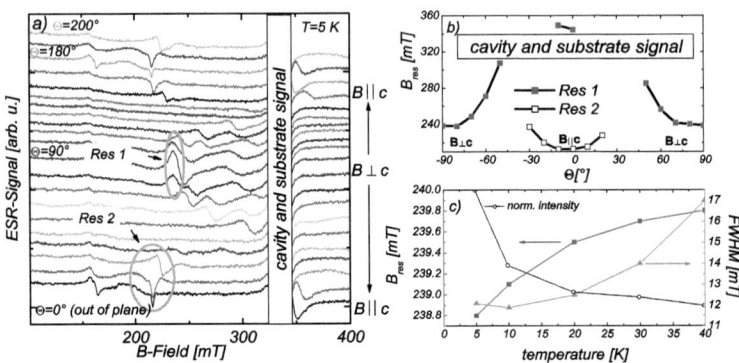

Figure 4.9: a) Stack plot of ESR-scans at different polar angles for a MBE grown sample with nominally $2 \cdot 10^{19}$ Gd /cm^3. b) shows the angular dependence of the two most prominent FMR-like lines visualizing the uniaxial behavior at least for Res 1. c) Resonance field (squares) and linewidth (FWHM, triangles) from Res 1 as a function of temperature. Above 40 K no resonance could be observed. Additionally the intensity of the resonance is plotted, normalized to the value at 5 K [102].

perature. The FWHM of 12 mT at 5 K increases up to 17 mT at 40 K. Such broad signals are typical for ferromagnetic resonance. Paramagnetic signals of this width - in particular, in a well ordered crystal environment at such low dopant concentration and temperatures of 40 K - are rarely known. With increasing temperature the resonance field shifts towards higher fields (g = 2 at 336 mT, ν = 9.4 GHz) but this effect is less evident than the line broadening - possibly due to the relative narrow temperature range. At 50 K the signal is already indistinguishable from the background. In addition the normalized intensity (the double integration of the shown resonance curves) is plotted in Fig 4.9 with respect to the 5 K data. While the FWHM and its temperature dependence of Res 1 point towards a ferromagnetic origin, the intensity resembles a paramagnetic, Curie-like 1/T behavior.

Further, a spin $\frac{1}{2}$-reference was used to estimate the absolute number of spins contributing to Res 1. The intensity in the out-of-plane geometry can be assigned to $2 \cdot 10^{13}$ 1/2-spins. The estimated absolute amount of Gd atoms in this sample is also

$2 \cdot 10^{13}$. Assuming that each Gd atom contributes with spin 7/2 the result indicates that the observed signal originates only from 1/7 of the Gd content. Therefore only a small fraction of the incorporated Gd is visible in ESR whereas the majority of Gd atoms is not.

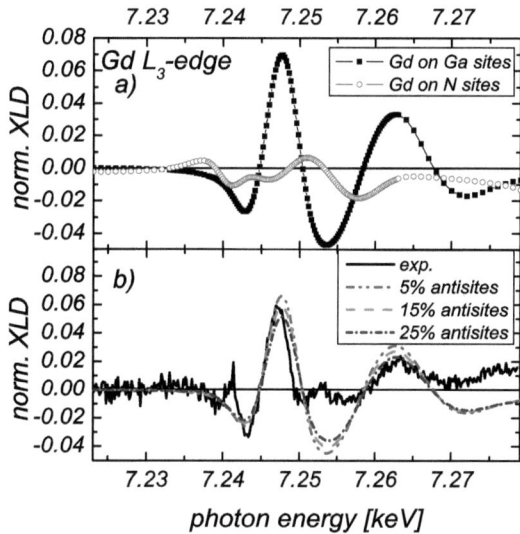

Figure 4.10: a) Simulated XLD spectra for Gd atoms on Ga- and N-sites, respectively. b) X-ray linear dichroism at the Gd L_3-edge proving the Gd to be predominately incorporated on substitutional sites. The largest deviations of experimental from simulated spectra occur at the muffin tin energy at 7.253 keV.

Element specific structural investigations by XLD yield further information on these two fractions. In figure 4.10 a) simulated XLD spectra for Gd on Ga and on N-sites are shown. The spectra were simulated with the FDMNES code for a $Ga_{23}GdN_{24}$ super cell. Details can be found in [134]. The two spectra in a) reveal very different XLD signals for substitutional and antisite incorporation of the dopant. The amplitude of the XLD signal of substitutional Gd is about seven times the signal of antisite Gd.

In b) the experimental XLD spectrum is shown. The curve is in good agreement with the expected spectral shape for substitutional Gd. Deviations around 7.252 keV originate predominantly from the muffin tin approximation of the simulation. Quantitative statements about the substitutional Gd incorporation could be derived from the amplitude of the main feature at 7.247 keV.

Additionally to the experimental data in 4.7 b) simulations for a different amount of antisites are shown. The simulated and experimental amplitude is fitted best for an amount of 15% of antisites. It is important to note that a random distribution of 15% of the Gd atoms would have a comparable effect of a reduced amplitude because these cause virtually no XLD signal.

Assuming 15% Gd-atoms in a different phase than paramagnetic Gd^{3+} on Ga sites, this result can be compared with magnetic resonance measurements, which have shown only 15% of the Gd to be source of the observed resonance signals. This would be consistent with the formation of ferromagnetic Gd or GdN clusters, which yield no XLD signal in case of a cubic phase or random orientation. Further, as shown in section 4.2 in figure 4.6, element specific hysteresis taken at T = 7 K and T = 40 K were not well-fittable with the same Brillouin or Langevin function. The origin of this magnetization behavior may be explained within a two phase model. While at T = 7 K and high magnetic fields (5 T) the paramagnetic Gd^{3+}-ions represent the dominating contribution to the magnetization, this changes at T = 40 K, due to the 1/T temperature dependence. At 40 K possible ferromagnetic clusters with a macrospin can represent a significant fraction of the Gd specific magnetization. Note that in figure 4.6 b) the measured XMCD signal is about $1 \cdot 10^{-2}$ larger than expected from the Brillouin fit. Assuming that at 7 K both phases could be almost saturated, whereas at 40 K this happens only for the clustered phase, the ratio of the different saturation magnetization contributions can be derived:

$$\frac{XMCD_{5T}(40K) - Brillouinfit_{5T}(40K)}{XMCD_{saturation}(7K)} \approx \frac{1 \cdot 10^{-2}}{6.5 \cdot 10^{-2}} = 0.15 \quad (4.3)$$

For further interpretation the XMCD data of figure 4.7 are reevaluated. The saturation magnetization is assumed to consist of two contributions:

$$M_s = 0.85 \cdot M_s(Gd^{3+}) + 0.15 \cdot M_s(cluster) \quad (4.4)$$

PHASE SEPARATION AND CLUSTERING

The paramagnetic, contribution of Gd^{3+}-ions shall strictly follow the Brillouin function:

$$M(Gd^{3+}) = M_s(Gd^{3+}) \times \underbrace{\left(\frac{2J+1}{2J} \coth\left(\frac{2J+1}{2J}y\right) - \frac{1}{2J} \coth\left(\frac{y}{2J}\right) \right)}_{B_J(y)} \quad (4.5)$$

with

$$y = \frac{g_J \mu_B J B}{k_B T} \quad (4.6)$$

The cluster contribution can be described by a Langevin function (which is the limiting case for $J \to \infty$ of the Brillouin function) for superparamagnetic particles:

$$M(cluster) = M_s(cluster) \times \underbrace{\left(\coth(y) - \frac{1}{y} \right)}_{L(y)} \quad (4.7)$$

with

$$y = \frac{mB}{k_B T} \quad (4.8)$$

In this case m represents the magnetic supermoment of the particle with arbitrary (infinite J) alignment with respect to an external field. In the following the nominator of y in the Langevin function is chosen to be 100 times the one of the Brillouin function ($m = 100 \cdot g_J \mu_B$) [3]. This simply reflects a higher magnetic moment (a supermoment) of a nanoparticle than a single ion. Note that such a nanoparticle has roughly a diameter of 2 nm [4], which is just below the detection limit of XRD. The larger nominator causes a different temperature dependence, since this dependency of Brillouin- and Langevin function is governed by $y \propto \frac{1}{k_B T}$. In figure 4.11 a) the black curve represents an Brillouin function with $J = S = 7/2$ which is close to saturation at high fields. Note that the XMCD data at 7 K of figure 4.6 show a similar behavior. The red and the green curves in figure 4.11 depict the reduced Brillouin function (85%) and the Langevin function which saturation magnetization reaches 15% of the black Brillouin function. The blue line represents the sum of the reduced Brillouin function and the Langevin function.

Already at this point it is worth to mention that the measured 7 K data in figure

---
[3] The actual size of m might vary over a wide range, since it hardly influences the following considerations as long as the Langevin function saturates at lower fields than the Brillouin function.
[4] Assuming a spherical particle with hcp structure.

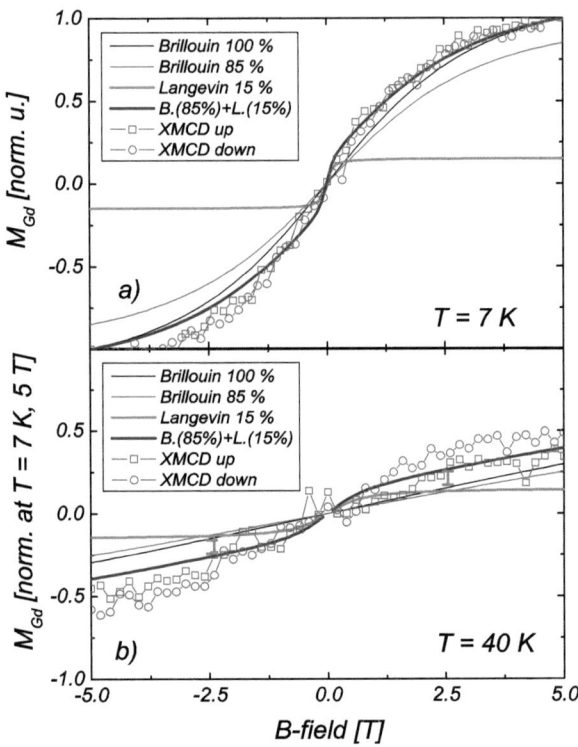

Figure 4.11: M(H) curves modeled by a superposition of Brillouin and Langevin functions. a) The saturation magnetization of a Brillouin function (black) is split into a 85% Brillouin like (red) and 15% Langevin like (green) magnetization. The sum of both contributions (blue) at T = 7 K still matches relatively well with the original 100% Brillouin function. b) At higher temperature the mismatch between both curves becomes obvious. Note that over a wide field range the difference corresponds approximately to the saturation magnetization of the Langevin function (orange bars). The rescaled XMCD data of figure 4.6 are plotted in addition (grey).

4.6 are underestimated by the Brillouin fit, which is adjusted to the saturation value of the XMCD data. In the simulated curves of figure 4.11 a similar mismatch can be seen between the black and blue curves. For better visualization the XMCD data of figure 4.6 have been normalized to the value at a field of 5 T. As can be seen in a) the superposition of the Brillouin - and Langevin function yields a better agreement with the experimental data.

For figure 4.11 b) Langevin and Brillouin function were calculated accordingly for $T = 40$ K and XMCD data were rescaled with the same factor as for the $T = 7$ K plot to obtain comparability. At this temperature the different behavior of the two phases becomes more pronounced since the slope of the Brillouin function is strongly reduced. In the available field range the Brillouin function shows a almost linear behavior, while the Langevin function still saturates at relatively low fields and acts solely like an offset at higher fields. Due to the small slope of the Brillouin function the absolute difference between the 100% curve and the 85% curve is rather small over a wide field range. Conclusively the difference (orange bars) to the superposition of Langevin and Brillouin function is given approximately by the saturation value of the Langevin function, which describes the clustered phase. The respective ratio of para- and superparamagnetic phase contribution was already calculated by equation 4.3. In contrast to the pure Brillouin- or Langevin functions used in figure 4.6 the weighted superposition in figure 4.11 also describes the experimental values at $T = 40$ K reasonably well.

A more realistic modeling of a cluster ensemble should account for size distributions, blocking temperatures and anisotropies as well a distribution of individual Curie-temperatures of the nanoparticles. However, the general mechanism of the superposition of a magnetic phase saturating at low fields and a pure paramagnetic one – hardy saturable at elevated temperatures – would remain the same. Therefore the temperature dependence of the element specific magnetization curve has to be considered as a clear evidence of the existence of phase separated clusters carrying supermoments in agreement with ESR and XLD results.

4.4.2 Gd:GaN with evidenced clusters of GdN

This section deals with a 2.9% Gd doped Gd:GaN sample grown on sapphire, where phase separated GdN can be detected by XRD. A analysis with the Scherrer formula

[138] yields a average cluster size of 27 nm [139]. Many results of this sample can be transferred to samples with lower dopant concentration to illustrate effects related to clustering.

The FC/ZFC curves of the 2.9% doped Gd:GaN sample are shown in figure 4.12. A clear separation of both curves up to approximately 60 K is visible. The low temperature regime is governed by the 1/T increase due to paramagnetic Gd. At higher temperatures up to 70 K the magnetization behavior of the sample is likely to result from the GdN phase. A mixture of superparamagnetic and blocked ferromagnetic cluster can account for the FC/ZFC splitting up to 70 K.

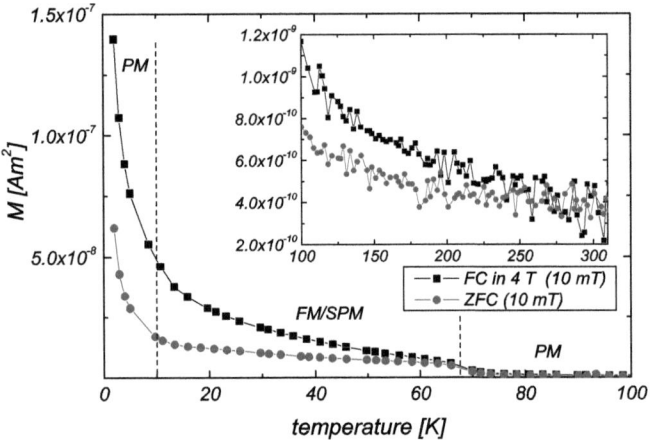

Figure 4.12: FC/ZFC curves indicating GdN phase separation. In the low temperature regime (< 10 K) paramagnetic contributions result in a 1/T behavior. Up to 70 K M(T) is governed by a SPM and FM phase. The sharp drop at 70 K reveals the breakdown of ferromagnetic order in GdN ($T_C(GdN) \approx 68$ K). The inset shows a separate measurement where the sample was cooled down only to 100 K. The residual FC/ZFC splitting indicates a very small amount of Gd cluster ($T_C(Gd) \approx 298$ K).

The FC and ZFC curves show a sharp drop at 70 K. This can be understood as the ferromagnetic to paramagnetic phase transition of GdN clusters big enough to show bulk like Curie-temperatures [5].

The inset of figure 4.12 shows the temperature range from 100 K to 300 K on an enlarged scale. Even in this temperature regime a small splitting of the measured magnetization curves is visible up to approximately 250 K. This indicates a residual amount of Gd clusters in this sample. However, the small signal is close to the detection limit [87].

Figure 4.13 shows magnetic resonance measurements in a) *out-of-plane* and b) *in-plane* geometry at different temperatures. Even though the signal-to-noise ratio is excellent the line shape is obviously not the derivative of neither a Gaussian nor a Lorentzian [6]. This is indicative for a superposition of resonance lines. Even though the signal is attributed to a distribution of intrinsic ferromagnetic GdN clusters a feature of bulk-like ferromagnets becomes visible: With increasing temperature the anisotropy is reduced resulting in a shift of the resonance field towards $g = 2$ ("$\frac{\omega}{\gamma}$") (≈ 337 mT). Conclusively the out-of-plane resonance field is reduced and the in-plane resonance field increased (see left insets). The angular dependency (upper inset in a)) of the line indicates relatively small anisotropies of the individual clusters. The black line is not a fit but a cosine function as guide to the eye. The angular dependence of paramagnetic centers in wurtzite structure ($B_{res} \propto \sqrt{g_{\parallel}^2 cos(\theta)^2 + g_{\perp}^2 sin(\theta)^2}^{-1}$) shows clear deviations (blue dotted curve). Thus a paramagnetic origin of the signal becomes unlikely.

The right inset of figure 4.13 b) shows the temperature dependence of the intensity, derived by the double integration of the signal. Already at 30 K the resonance is difficult to be identified because of the background signal as indicated by the hatched area in the plot.

In figure 4.14 the dependence of the intensity of the ESR signal on the incident microwave power in the out-of-plane geometry is shown. In b) and c) the corresponding spectra are plotted, whereas a) shows the peak-to-peak amplitude of the signal which is proportional to the intensity in case of constant line shape. The measurement was taken with fixed spectrometer settings to ensure comparability. The power dependency is expected to be $I_{pp} \propto \sqrt{P_{MW}}$, since only magnetic (dipole) transitions cause the energy absorption.

---

[5]T_C(GdN)≈ 68 K

[6]The feature appearing with increasing temperature at low fields in a) was assigned to a sample holder contamination.

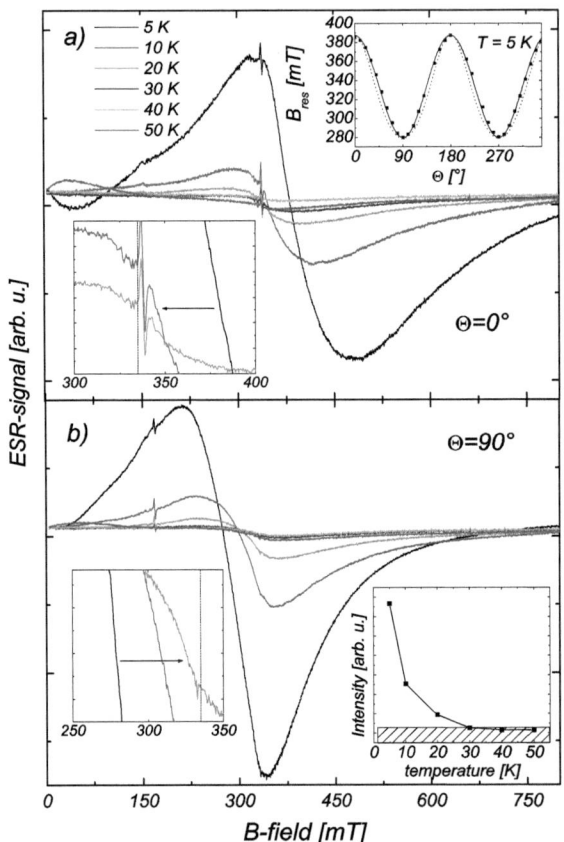

Figure 4.13: Temperature dependence of the resonance signal for in-plane and out of plane geometry for clustered Gd:GaN ($\nu = 9.4$ GHz). A reduction of the anisotropy with increasing temperature manifests in shifts of the resonance field towards g = 2 ("$\frac{\omega}{\gamma}$") for in/out-of-plane orientation. The narrow superimposed line is a contribution from the sapphire substrate. The upper inset in a) shows the angular dependence of the resonance. In the right inset in b) the temperature dependence of the signal intensity is plotted.

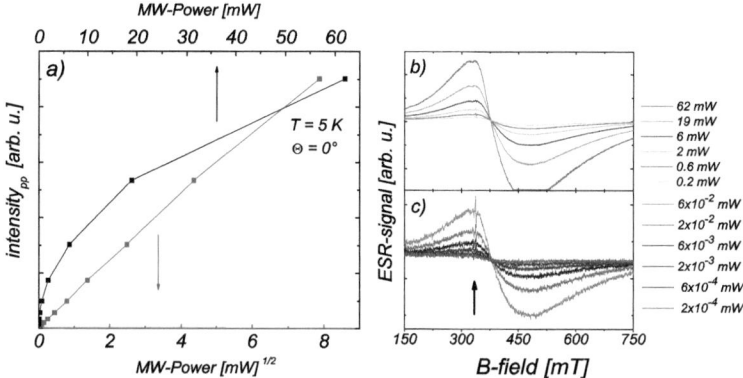

Figure 4.14: Power dependence of the signal intensity. A monotonic behavior with increasing power is visible. Note that the narrow line (black arrow in c)) can be saturated as it vanishes at high power values.

For paramagnetic resonance an important deviation from this proportionality occurs in case of saturation, which means an equal population of the respective energy levels. In this case the induced emission of photons cancel out the absorption and the resonance line vanishes (see e. g. figure 3.7). This is hardly possible in ferromagnetic resonance, since the photon interacts with an exchange coupled spin system. The latter is also valid for the magnetic moment of single domain nanoparticles, at the available microwave powers. Figure 4.14 a) shows a monotonic increase of the signal intensity with increasing power, which is therefore consistent with a superparamagnetic cluster ensemble. Note that the paramagnetic impurities of the substrate are saturated at high power values (black arrow in figure 4.14 b) and c)).

In addition to the integral measurement methods, like SQUID or ESR, element specific investigations were conducted for the clustered sample and are summarized in figure 4.15. The black curve in a) shows the XANES at the Gd L_3-edge. The high Gd concentration and the layer thickness ($\approx$ 500 nm) result in a clear absorption edge. The respective XMCD curves have a high signal to noise ratio. At 150 K the dichroic signal is still sufficient large for hysteresis measurements. However the 300 K XMCD data do not indicate any magnetic dichroism of the dopant. This is

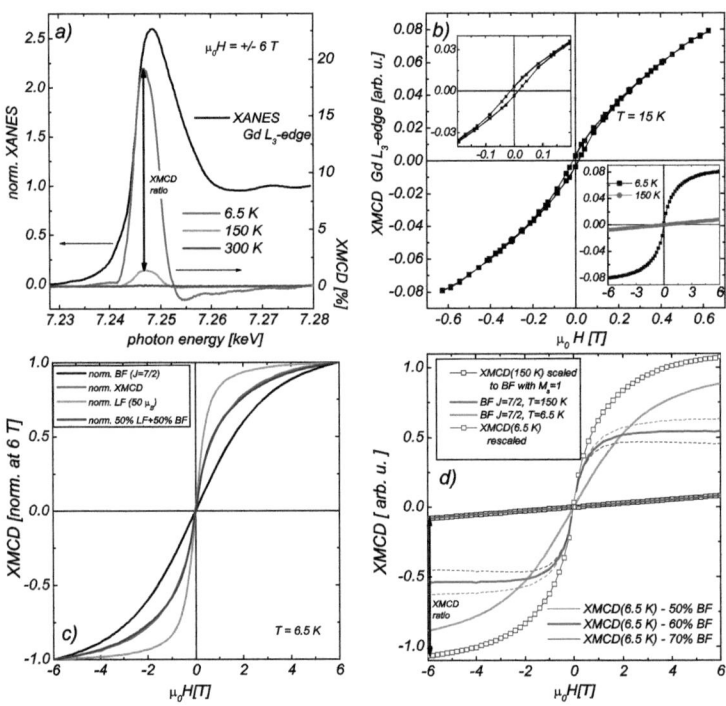

Figure 4.15: Synchrotron measurements on clustered Gd:GaN. a) Gd L_3 absorption edge spectrum and XMCD at different temperatures. b) Element specific M(H) curves. The upper inset shows the opening of the hysteresis. The lower inset shows the saturation at high fields. c) Fitting of the experimental data by superposition of a Langevin and a Brillouin function. d) Evaluation of a *non-Brillouin-like* M(H) component (see text).

PHASE SEPARATION AND CLUSTERING   93

in agreement with a paramagnetic behavior leading to a negligible magnetization at high temperatures.

In figure 4.15 b) the XMCD hysteresis measured at 15 K over a field range from -0.6 to +0.6 T with high field resolution is shown. Besides an overall s-shape a remanence and coercivity is visible. The upper inset in b) shows the open hysteresis on an enlarged scale. Two additional XMCD hysteresis are shown in the lower inset of b) at temperatures of 150 K and 6.5 K, respectively. These measurements cover a ten times larger field range (6 T) with less field resolution, showing the saturation behavior. Further analysis of these data was carried out in c) and d) which will be described in more detail in the following:

The sample is assumed to consist of two contributions, paramagnetic Gd^{3+}-ions and a population of superparamagnetic clusters, in particular Gd or GdN agglomerations [7], which was already discussed along with the SQUID measurements in figure 4.12.

$$M_s = x \cdot M_s(Gd^{3+}) + (1-x) \cdot M_s(cluster) \qquad (4.9)$$

One approach to separate the different magnetization contributions relies on the XMCD M(H) curves. For this, the M(H) curve is normalized at a field value of 6 T (figure 4.15 c)). The respective Brillouin function is normalized in a similar way. Obviously the Brillouin function cannot describe the experimental data. Using the ansatz 4.9 the experimental data are fitted best for a superposition of 50% of a Langevin function (m = 50 μ_B) and of a 50% Brillouin function (J = S = 7/2). Note that smaller supermoments result in deviations at higher fields and larger supermoments in deviations at small fields which cannot be adjusted by changing fractions of Langevin or Brillouin like components, respectively.

A second possibility to evaluate the data is illustrated by figure 4.15 d). The XMCD at 150 K data are rescaled to a Brillouin function (red curve) with normalized saturation magnetization, since at 150 K all Gd atoms assumed to behave paramagnetic. The experimental data taken at 6.5 K are rescaled according to the XMCD ratio shown in figure 4.15 a). Obviously the experimental data (blue squares) show a different magnetization than expected from the Brillouin function at 6.5 K (green curve).

---

[7]This approach neglects the small remanence shown in b) but since conclusions are solely derived from high field data the deviations are acceptable.

To extract the magnetic behavior of the non-Brillouin like component, the Brillouin function is subtracted from the experimental data. The residual M(H) curve shows a negative slope with increasing field values (not shown), which is unphysical. This can be adjusted by assuming less Gd atoms to behave paramagnetic at low temperatures, i. e. the weight of the Brillouin function is reduced. Assuming full saturation of the residual M(H) component within the magnetic field range, the Brillouin component has to be reduced to 60%. In turn this means a 40 % Langevin component (pink curve) according to the ansatz in 4.9. Additionally the residual M(H) curves for a 50% and 70% Brillouin contribution are plotted (dotted lines), showing slopes different from zero at high fields.

Finally, the different results of the two evaluation methods have to be discussed.

The assumption of a fully saturated residual M(H) component might not be fulfilled. A distribution of supermoments of different size in the sample seems more realistic. In this case very small supermoments will not be fully saturated at high fields. Taking this into account, the non-Brillouin like M(H) curve might still have a positive slope at high fields. Therefore the alternatively shown M(H) curve in d) for 50% (orange), could also be reasonable. Thus the different derived fractions of both methods can agree within the error bar.

The second analysis method crucially depends on the XMCD ratio derived from a), which means another problem has to be accounted for, which is visible in figure 4.15 d): The experimental data in d) have been scaled to the normalized Brillouin function at 150 K, therefore the low temperature data, rescaled by the XMCD ratio, should also be normalized. Instead the data clearly exceed the saturation value of 1. Assuming a solely paramagnetic Gd^{3+} contribution at 150 K, the ratio of the magnetization at 150 K to the magnetization at 6.5 K for a applied field of 6 T can be calculated. The respective Brillouin function is based on $J = S = 7/2$ and $g = 2$. The derived ratio can be compared with the ratio of the maximum XMCD values [8].

$$\frac{B_{7/2}(150K, 6T)}{B_{7/2}(6.5K, 6T)} = 0.0909 \neq \frac{XMCD(150K, 6T)}{XMCD(6.5K, 6T)} = 0.0746 \quad (4.10)$$

This result is indicative of some excess magnetization as measured by XMCD at 6.5 K, since at 150 K all Gd^{3+} ions are assumed to behave paramagnetic. The additional signal can be calculated:

$$\frac{B_{7/2}(150K)}{B_{7/2}(6.5K)} = 0.0909 = \frac{XMCD(150K)}{XMCD(6.5K) - x} \rightarrow \frac{x}{XMCD(6.5K)} \approx 18\% \quad (4.11)$$

---

[8]A constant magnetic moment per Gd atom is assumed.

As will be shown in the following this XMCD fraction cannot be explained by a ferromagnetic phase transition, since also the paramagnetic phase is close to saturation at 6 T. A superparamagnetic phase at 6.5 K and 6 T is approximated to be fully saturated. Therefore the low temperature signal will consist of a paramagnetic phase (fraction a) which is 88% saturated ($B_{7/2}(6.5K, 6T)$) and a 100% saturated superparamagnetic phase (fraction $1 - a$):

$$\frac{B_{7/2}(150K)}{a \cdot B_{7/2}(6.5K) + (1-a) \cdot 1} = 0.0746 \Rightarrow a = -0.64 \qquad (4.12)$$

This result means a fraction of only $a/(1-a) \approx 39\%$ of paramagnetic Gd atoms at T = 6.5 K. However in view of the negative sign of a the ansatz in 4.9 has to be questioned, since it means that the sample has more magnetic moment at 6.5 K than at 150 K.

This can also be interpreted the other way round, resulting in a missing magnetic moment at 150 K, which can be determined to be 22%.

Antiferromagnetic coupling and therefore loss of magnetic moment can be considered as explanation. Antiferromagnetic phases of rare earths compounds are well known and subject to recent research [140]. In particular the separation of a low temperature ferromagnetic phase from a paramagnetic phase by an antiferromagnetic phase at elevated temperatures has been shown [141]. However, GdN has been considered as an exception since such a separation is so far unknown. Nevertheless the special conditions of nanoparticles embedded in a host crystal might easily alter the effective magnetic coupling resulting from competing RKKY and superexchange interactions [142]. Note that the magnetic moment used for the Langevin component in figure 4.15 c) is very small with respect to the XRD estimated cluster size of 28 nm, also indicating the compensation of moments.

In summary these considerations show that both evaluation methods used to derive the fraction of a superparamagnetic phase agree within the error bar. This is a strong evidence for a clustering tendency in Gd:GaN at a concentration of 2.9% Gd.

4.5 Discussion: Gd:GaN

The dilute magnetic semiconductor Gd:GaN, which is claimed to show ferromagnetism even far above room temperature, has been studied in concentrations varying from $10^{-5}\%$ to 2.9% of Gd atoms. The low concentration regime could be addressed by focused ion beam implantation whereas higher concentrations were obtained by MBE growth.

Basically two sets of hexagonal GaN samples implanted with Gd were under investigation. The sets were grown on different substrates, 6H-SiC(0001) and sapphire (0001), but with comparable high crystalline quality as shown by XRD. Structural effects of the implantation were only observed for highest implantation doses of $1 \cdot 10^{15}$ Gd/cm^2 corresponding to concentrations of $1 \cdot 10^{20}$ Gd/cm^3, respectively.

Magnetic characterization by SQUID measurements yields differing results for both types of samples. While the Gd:GaN films grown on SiC exhibits ferromagnetic-like features even at room temperature, the series of films grown on sapphire shows no indications of ferromagnetism.

In the latter series a 20 mT broad magnetic resonance could be observed, which roughly scales with the implantation dose. No angular dependency and a g-value of 2.00 was found, which is typical for a 4f^7 ion in the $S_{7/2}$ state. The structural damage evidenced by XRD and the absence of any fine structure in the ESR spectra on the contrary point towards paramagnetic defects induced by implantation as origin of the resonance. This is corroborated by the magnetization values as measured by SQUID at low temperatures and high fields which exceeds the possible Gd contribution.

A similar magnetic resonance signal was also found in the highest implanted sample grown on SiC accompanied by additional paramagnetic signatures. All these magnetic resonances vanish after rapid thermal annealing proving no causality between the room temperature ferromagnetic-like behavior and these signals, since this is unchanged. The absence of any Gd^{3+} resonance signature is probably caused by high fractions of non substitutional ions and line broadening effects due to reduced structural quality after implantation.

Element specific information of the role of Gd in Gd:GaN was yielded by synchrotron measurements on MBE grown samples. XMCD-M(H) curves prove a non-hysteretic behavior of Gd, also for samples showing a room temperature hysteresis when measured by SQUID.

Investigation of the local structural condition by XLD revealed that Gd is likely to be incorporated on Ga-sites as Gd^{3+} ion (85% XLD amplitude of the expected XLD

DISCUSSION: Gd:GaN

signal for substituional Gd^{3+}) in MBE grown samples [9]. In a sample which shows good structural quality by XRD and XLD as well as SQUID room temperature hysteresis, the XMCD M(H) curve revealed a superposition of super- and paramagnetic behavior of the dopant. Experimental M(H) were modeled using Brillouin and Langevin functions to determine the fractions of the clustered, superparamagnetic and the purely paramagnetic phase. A ratio of clustered to paramagnetic phase of 1/7 was independently verified by magnetic resonance measurements. These findings clearly indicate secondary phases as origin of ferromagnetic signatures in Gd:GaN. XMCD measurements performed at the Ga K-edge in two different sample geometries revealed a tiny polarization of the host cation. This polarization is shown to be at least one order of magnitude too small as needed for models claiming a long range magnetic ordering in Gd:GaN caused by polarized spheres around the dopant. If one disregards SQUID results, Gd:GaN has to be considered as a (super-) paramagnetic material. Ferromagnetic-like behavior up to and above room temperature as found in some samples when measured by SQUID, might be caused by particularities of the GaN crystals not taken into account so far. Note that paramagnetism has also been found for Gd implanted cubic GaN [124]. However, possible mechanisms of a ferromagnetic behavior have to be beyond what is suggested in literature. In view of the intense research efforts of different research groups and the poor reproducibility of room temperature ferromagnetism in Gd:GaN, ferromagnetic contaminations have to be considered.

The growth of Gd:GaN always contains the risk of phase separation. The effects of Gd and GdN clusters on XMCD and FMR measurements are discussed. Results of element specific hysteresis and magnetic resonance can be understood within the framework of cluster distributions of different T_C.

These results stress the importance of considering phase separation as an origin of ferromagnetic signatures at low temperatures in integral measurements like SQUID-magnetometry. Ferromagnetism at and above T = 300 K can not be explained, neither by a GdN phase nor a Gd phase.

---

[9]Note that for the implanted samples the Gd concentration was too low to be measured by XLD.

Chapter 5

Experimental results for Co:ZnO

5.1 Preparation of Co:ZnO samples

In contrast to Gd:GaN the portfolio of Co:ZnO samples available for this work was much more diverse, since this material was also self grown by reactive magnetron sputtering (RMS) as described in section 2.1. Similar to GaN the low doping regime was realized by FIB implantation. Samples with higher concentrations were available from three common epitaxial growth techniques, namely RMS, PLD and MBE. Co:ZnO samples used for this work were provided by the following collaborators:

- Samples grown by PLD on sapphire were provided by the group of Prof. S. A. Chambers [1]. Details of the growth process can be found e.g. in [143].

- One sample grown by PLD but on r-plane sapphire and at the very high growth temperature of 600 °C was provided by the group of Prof. H. Adrian [2].

- High dilute samples were produced by focus ion beam implantation of Co^+-ions into ZnO substrates from the CrysTec company [97]. All implantations were conducted by the group of Prof. A. D. Wieck [3].

- Nanocrystalline Co-doped ZnO powders grown by the group of Prof. Winterer [4] were available for investigations. The powders were fabricated by Chemical Vapor Synthesis. Details can be found in [78] and [144].

[1] Pacific Northwest National Laboratory, Richland, Washington 99354, USA
[2] Institute of Physics, University of Mainz, Germany
[3] Lehrstuhl für angewandte Festkörperphysik, Ruhr-Universität Bochum
[4] University of Duisburg-Essen, Lotharstrasse 1, D-47057 Duisburg, Germany

Appendix B provides a tabulated overview of the most important samples, the results of which will be shown in the following sections.

5.2 Co:ZnO - samples of high structural quality

In this section results of Co:ZnO samples of highest structural quality, as measured by XRD and XLD are compiled. For high dopant concentrations RMS and PLD grown samples were under investigation. Consistent results of ion implanted samples with low dopant concentrations are briefly mentioned at the end.

5.2.1 ZnO doped with 10% Co

Very detailed experiments have been conducted on PLD and RMS grown Co:ZnO samples with a nominal dopant concentration of 10%. For one PLD grown sample the dopant concentration has been determined by PIXE. The measured Co content was 10.8%. The concentration of the RMS samples was given by the composition of the sputter targets. Deviations from the ratio may occur, since sputtering rates differ for the respective elements. The composition of one RMS grown sample was also determined by PIXE yielding a slightly reduced Co content of 9.5%. The high structural quality of PLD grown samples has also been achieved by RMS. The latter method was available within the group, which enabled testing of various growth parameters and has lead to a thorough control of the structural quality. Details of the growth process are given in section 2.1. The following section is focused on 10% doped Co:ZnO samples with high structural quality.

XRD 2θ-scans with a FWHM of $0.15°$ at the ZnO (002) peak are reached for RMS and PLD grown samples, proving the excellent structural quality. Besides the standard characterization by XRD the local structural quality of the cations was investigated by XAS and XLD.

Figure 5.1 compiles a complete analysis of the local structure of the Zn- and Co-cation sites of a PLD grown sample (SC042807B). In a) and b) the Zn and Co K-edge absorption spectra for vertical and horizontal linear polarized light are shown, respectively. The c-axis of the sample is chosen to be parallel to the vertical polarization direction. All four spectra show a pronounced fine structure of the 1s→4p absorption. For higher energies the spectra continue into the EXAFS range. Differences in the fine structure for the two linear polarizations are visible for both cation

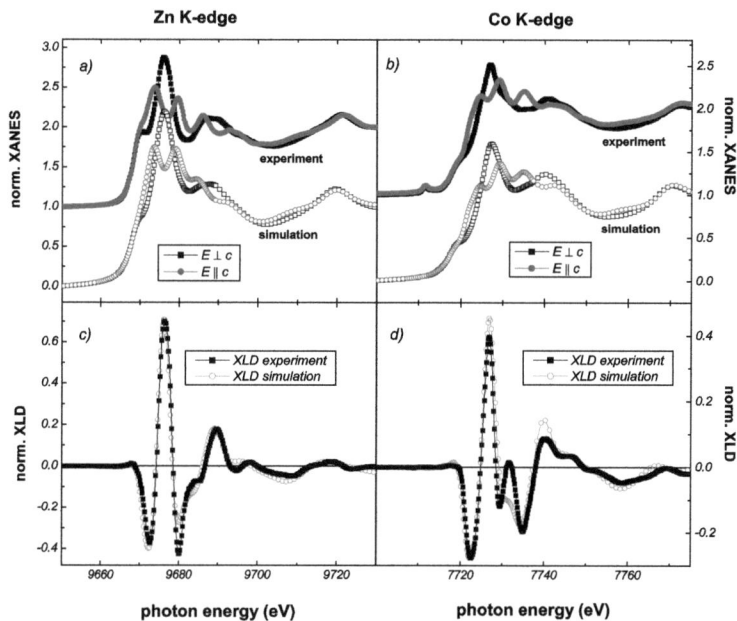

Figure 5.1: Structural analysis of PLD grown Co:ZnO by XAS and XLD. a) and b) show the absorption Zn- and Co K-edge for vertical and horizontal linear polarized light, respectively (full symbols). Corresponding simulated curves are plotted as open symbols to show the very good agreement of results by the FDMNES code. c) and d) show the respective difference of the XANES spectra - the XLD. From the amplitude of the XLD of the Co K-edge compared to the simulation one can estimate over 95% of substitutionaly incorporated Co atoms - proving the high structural quality [24].

types.

In case of the Co K-edge absorption spectra a special feature is noteworthy. Before the actual absorption edge a small peak is visible, almost of similar size for both polarization directions. This *preedge feature* is known to be characteristic for Co^{2+}. The origin are hybridized states of the 3d and 4p orbitals which are lower in energy than the 4p final states of the absorption edge [145].

Below the experimental spectra in a) and b) simulations of the respective spectra are shown, calculated with the FDMNES code as mentioned in section 2.4. The agreement of the computed spectra with the shape of the experimental curves is generally very good, despite deviations caused by the muffin-tin approximation at the energies of 9680 eV and 7740 eV, for Zn and Co, respectively. The qualitative comparison with the shape of the experimental curve yields an important additional crosscheck of the structural quality besides XRD, since the crystal structure for the simulations is part of the input file. Note that the *preedge feature* is not accounted for by the simulations.

In the following sections XANES spectra are shown as isotropic (powder) spectra for better visibility, i.e. no individual spectra for I_{ver} and I_{hor}, since the information gain from these is contained in the XLD.

The lower plots c) and d) in figure 5.1 depict the difference of the absorption spectra for both polarizations, namely the XLD and the equivalent derived from the simulations. Both cations show the signature typical for wurtzite. All absorption spectra are normalized with respect to absorption values before and after the edge at the same energy. Therefore the derived XLD amplitudes [5] of different samples can be compared quantitatively. Simulation parameters were optimized for the spectra taken at the Zn K-edge of the best samples. A comparison of the experimental and computed XLD of Co-atoms on Zn-sites reveal that in the respective sample at least 95% of the Co atoms are incorporated on substitutional Zn-sites. Co atoms on different sites should result in altered XLD whereas Co atoms on random positions inevitable reduce the amplitude of the XLD signature [6]. This result corroborates the excellent structural quality on a local atomic scale as suggested by XRD.

In Figure 5.2 the structural characterization of a sputtered sample (080428) in

---

[5]The XLD amplitude refers to the first minimum and maximum of the spectrum.

[6]Extensive studies of possible Co positions and their XANES and XLD spectra have been performed and will be published. Some details can be found in [93].

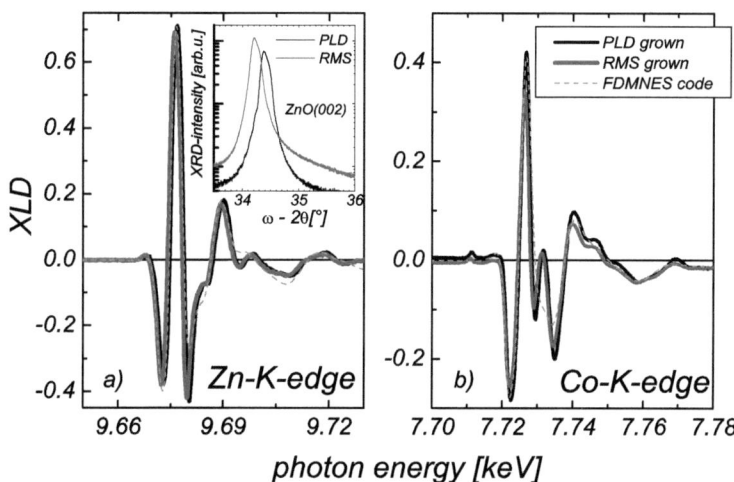

Figure 5.2: XLD measurements at the Co and Zn K-edge of high quality Co:ZnO sample grown by RMS in comparison to the PLD grown sample. While the Zn K-edge in a) reveals a XLD amplitude almost similar to the PLD sample and the simulation, the XLD at the Co K-edge in b) is reduced to 84%. Therefore less Co atoms have proper wurtzite environment tantamount to less structural perfection on the local scale [146]. The inset in a) shows the ZnO (002) peak of the XRD scan proving comparable structural quality on a global scale of both samples.

comparison to the above mentioned PLD grown sample (SC042807B) is compiled. The amplitude of the XLD spectra at the Zn K-edge is hardly reduced; which is shown in a). The XLD amplitude of the dopant in turn is reduced to about 84% in comparison to the PLD sample as can be seen in b). Therefore a reduced amount of Co atoms are in proper wurtzite environment, i.e. on substitutional sites.

The inset of figure 5.2 a) presents the corresponding XRD measurements which corroborate the comparable structural quality of the sputtered sample on a global structural scale. The small difference of the peak positions indicate a 4‰ smaller c lattice constant in the PLD sample, which might be due to a slightly different Co concentration.

Even though the dopant is less good incorporated in the sputtered sample, it has all magnetic signatures which are characteristic of high quality Co:ZnO, as will be shown in the following.

The Co specific magnetic polarization of the PLD sample is plotted in figure 5.3 as measured by means of XMCD. A comparison with the absorption curve which is also plotted shows that the most intense dichroic signal is observed at the preedge feature with a value of 0.3% (black arrow) at an energy of 7712.16 eV.

This energy value was taken for M(H) measurements as presented in the lower graph. Magnetization curves were taken for both field directions (solid/open symbols) over a range of 6 T at two temperatures of 6.7 K and 40.5 K, respectively. Both M(H) curves indicate no ferromagnetic behavior. The curve taken at 6.7 K reveals an s-shape without remanence, indicating only paramagnetic contributions from the cobalt atoms in the sample. At 40.5 K the hysteresis resembles a straight line with maximum magnetization values for highest fields much below the 6.7 K hysteresis. The functional dependency from applied field and temperature is well described by a Brillouin function assuming $S=3/2$ and $L=1.07$. These values correspond to a L/S ratio of 0.7 derived by application of the sum rules in [147] for a PLD grown 10% doped Co:ZnO sample. The respective magnetic moment per Co atom is $\approx 4~\mu_B$.

Figure 5.4 shows integral measurements of the magnetization behavior as measured by SQUID for high structural quality 10% Co doped Co:ZnO grown either by PLD or RMS. The curves are adjusted to the same magnetization value at 4 T for better comparison. Obviously the M(H) shapes are similar, independent of the growth method. Adjusting a Brillouin function - again assuming $S=3/2$ and $L=1.07$, leads to a very good agreement with the experimental data.

The presented measurements give a comprehensive, integral and element specific view of the magnetic properties of 10% Co doped ZnO of high structural quality. They yield no indications for ferromagnetism down to temperatures of 5 K. Room temperature SQUID measurements conclusively do not show any indications of ferromagnetic behavior (not shown). The presented results on $Co_{0.1}Zn_{0.9}O$ so far are astonishing, since theoretical predictions and experimental results are reported which claim strong interactions between the Co atoms resulting in long range magnetic order at this Co concentration [21, 27].

What has been not mentioned so far is a misfit between the paramagnetic behavior

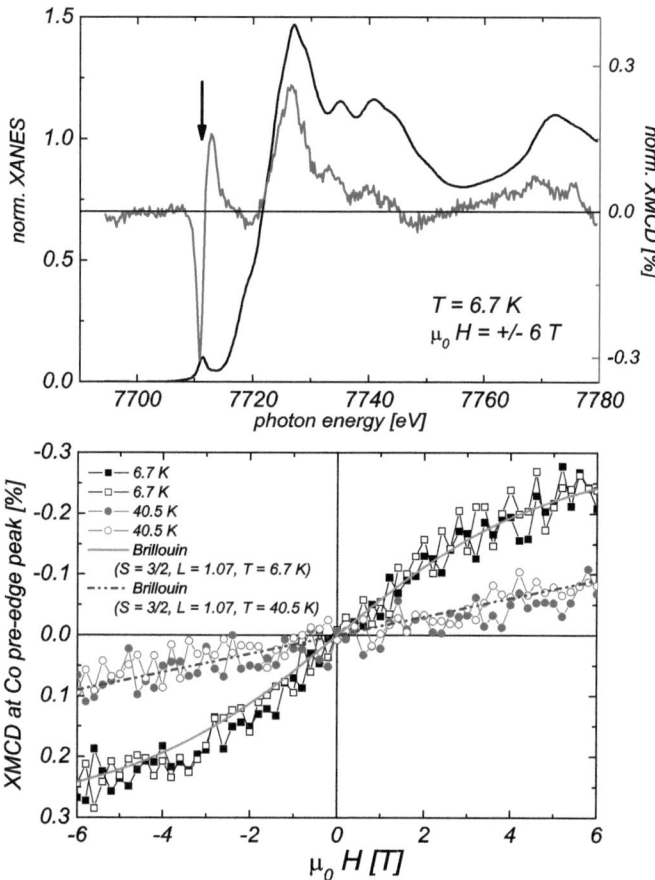

Figure 5.3: XMCD at the Co K-edge and hysteresis taken at the Co preedge feature. The element specific hysteresis curves were measured at two distinct temperatures 6.5 K and 40.5 K. Both magnetization curves are well fitted by the Brillouin function with S=3/2 and L=1.07 [24].

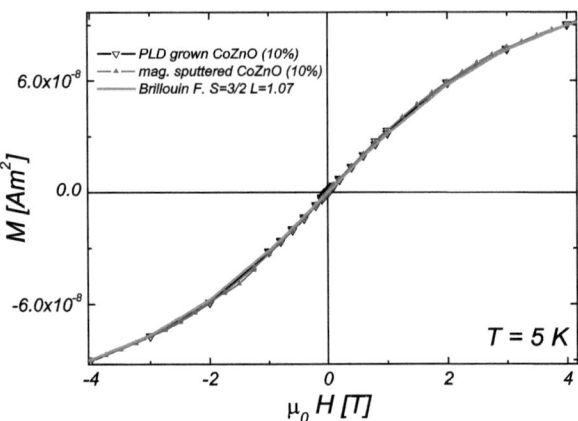

Figure 5.4: Magnetization curves measured by SQUID for best grown 10% Co doped Co:ZnO. Independent of growth method the magnetization can be described by a Brillouin fit. The value of L=1.07 fulfills the L/S ratio of 0.7 as reported in literature [147].

of Co, its magnetic moment and the absolute magnetization as measured by SQUID. With sample size, dopant concentration and thickness, the expected saturation magnetization for a film containing paramagnetic Co^{2+}-ions can be calculated. For this purpose the measured moment resulting only from the z-component of J has to be taken into account, i.e. μ_z. Additionally one has to consider measurement conditions as given by the SQUID, in particular a external field of 5 T and a temperature of 5 K.

$$\begin{aligned}\mu(Co^{2+}) = 4.8 \ \mu_B \ &\rightarrow \ \mu_z(Co^{2+}) = 4.1 \ \mu_B \ \rightarrow \\ M(T=5 \ K, \ \mu_0 H = 5 \ T) &\approx \ 0.88 M_{sat}(T=0 \ K)\end{aligned} \quad (5.1)$$

For the high structural quality PLD sample discussed before the calculation of the samples magnetization yields $(3.6 \pm 0.3) \cdot 10^{-7}$ Am^2 in stark contrast to $1 \cdot 10^{-7}$ Am^2 measured by SQUID. The SQUID and the XMCD hysteresis are both well described by a Brillouin function, thus the remaining fraction of Co has to be magnetically inactive.

Co:ZnO - SAMPLES OF HIGH STRUCTURAL QUALITY

To understand this discrepancy the statistical abundance of different Co atom configurations has to be considered. In chapter 1 the calculated probabilities according to the work of Behringer [66] were already introduced. The respective formulas for a 10% statistically distributed Co doping yield only a 28% probability to find a Co atom without any other Co atom as a next cation neighbor. Therefore the value of the total magnetization can be well understood considering only the magnetic moments of the single Co atoms contributing to the M(H) curve. Pairs of Co-O-Co occurring with a probability of 18% are thus likely to fully compensate their magnetic moments. Co triple configurations or higher order might contain partly frustrated magnetic moments, leading to a small contribution for the total magnetization. Different scenarios of possible additional magnetic contributions are discussed in [24], but result at maximum in a 8% fraction of the maximal Co induced magnetization.

As shown before one can expect Co^{2+}-ions on substitutional cation sites to be the governing magnetic contribution in high quality 10% doped Co:ZnO. Consequently magnetic resonance measurements could be expected to detect the well known spectra of this paramagnetic center. Despite thorough and extensive examination hardly any signals could be found in high quality samples of $Co_{0.1}Zn_{0.9}O$.

The reason for this are enhanced interactions between the ions which lead inevitably to a perturbation of the electron states addressed by ESR. These are most likely to be long range dipolar interactions.

The broadening of a paramagnetic signal with increasing concentration of the dopant is shown exemplarily for Co:ZnO nanopowders in figure 5.5. The samples were kindly provided by the group of Prof. Winterer [7]. Although the ESR spectrum of a powder differs from the spectra of epitaxial films (see section 1.3.2), the broadening can be well observed. In figure 5.5 spectra for three different concentrations of Co are shown. The two lower concentrations were verified by atomic absorption spectroscopy (AAS) whereas the third one was extrapolated.

Figure 5.5 a) shows the same data as in the lower plot b). In a) the spectra are adjusted by the spectral feature of the cavity around 330 mT. In this plot the Co^{2+} signature of the powder with the highest concentration is hardly visible. In 5.5 b) the respective spectra are scaled to the same height of the Co^{2+} signature. This lower plot visualizes the line broadening. Note that already the powder with 2.8%

---

[7]University of Duisburg-Essen, Lotharstrasse 1, D-47057 Duisburg, Germany

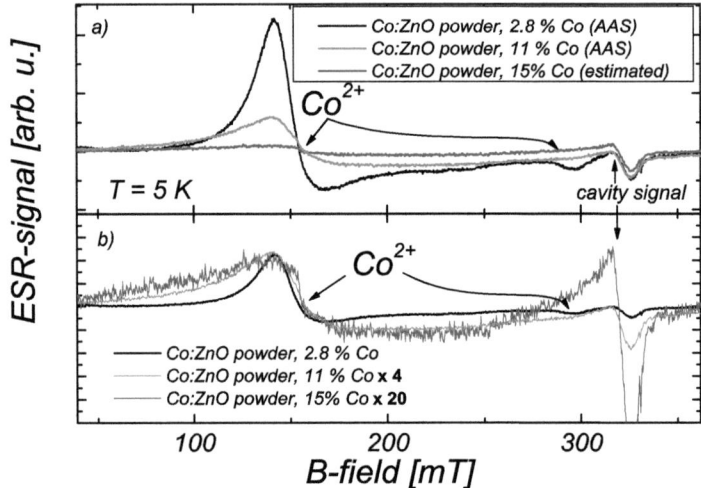

Figure 5.5: Line broadening observed in Co:ZnO nanopowders. Both plots show the same measurements. In a) the spectra are rescaled with respect to the cavity signal. In b) the spectra are rescaled with respect to the Co^{2+} related resonance signal.

of Co shows an intense line broadening.

A thorough treatise of the change of linewidth with dopant concentration due to exchange and dipolar interaction can be found in [69, 148]. Taking into account these interactions of paramagnetic centers in a crystal it is shown that dipolar interaction result in a line broadening by calculation of the second moment, $<\Delta\nu^2>$, of the frequency. Exchange coupling in turn results in contributions to the fourth moment, $<\Delta\nu^4>$, of the frequency which causes a pronounced peaking of the absorption curve, i.e. so-called exchange narrowing. A recent study on MBE grown Co:ZnO suggest contribution from both, exchange and dipolar, effects to the line shape [111].

As introduced in section 2.2 the double integrated resonance signal is proportional to the $M \cdot V$. Therefore it is intuitive that intense line broadening leads to a vanish-

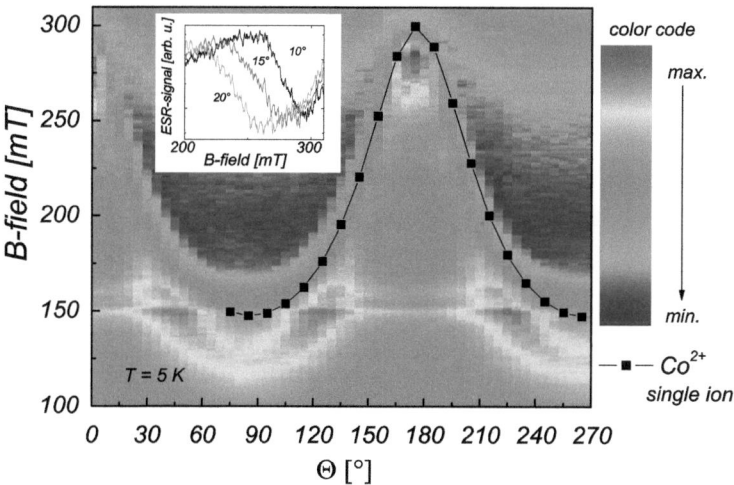

Figure 5.6: Color plot of angular dependent ESR signals in 5% Co doped ZnO. The colors are assigned according to the ESR signal. The angle $\theta = 0°$ corresponds to $\vec{B} \parallel \vec{c}$. A uniaxial lineshift within the 270°-scan is clearly visible. The inset shows exemplarily the ESR resonance signals for 10°, 15°, and 20°. The black symbols are measured resonance values of the Co^{2+} single ion spectrum.

ing signal. To clarify whether the RMS grown Co:ZnO samples contain a extremely broadened Co^{2+}-ion signal, a sample optimized for ESR measurements was grown. The thickness of the film was increased by the factor of ten (≈ 1 μm) and the dopant concentration was reduced to 5%. The higher amount of material results in an increased total signal size, whereas the lower dopant concentration reduces the line broadening.

Figure 5.6 shows the result of ESR scans of an angular range from 0° to 270° at 5 K. The maximum value of the ESR signal corresponds to red color, minima are visualized by blue regions. The color plot taken with a resolution of 5° reveals the shift of a broad resonance of about 75 mT linewidth. In the inset the ESR signals for three angles are exemplarily shown. The black squares represent the resonance fields of the Co^{2+} spectrum in case of highly dilute isolated single ions as it is dis-

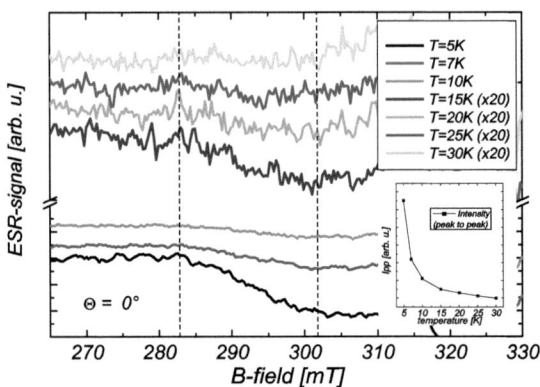

Figure 5.7: Vanishing of the resonance line with increasing temperature in 5% doped Co:ZnO. The black lines indicate roughly the line position. Within the small temperature range no shifting can be observed. Note: A slight slope of the background has not been corrected.

cussed in detail in chapter 3. Within the accuracy of the measurement the resonance fields, and thus the g-values, of the observed broad line and the Co^{2+} single ion are identical.

Even though the resonance line in this 5% doped sample shifts angular dependent similar to the Co^{2+}-ion the linewidth might also suggest a ferromagnetic origin. Besides the purely paramagnetic behavior observed by SQUID for this sample (not shown), a ferromagnetic origin it can be also shown by magnetic resonance that a ferromagnetic origin is very unlikely. Figure 5.7 shows temperature dependent measurements of the line. The amplitude of the ESR-signal vanishes rapidly with temperature like $1/T$. At 40 K a possible line is no longer distinguishable from the background of the measurement. The resonance fields of ferromagnetic signals typically shift with temperature. The vertical dotted lines in the graph indicate no such shift. Therefore also magnetic resonance corroborate a paramagnetic origin of the signal.

Considering the ESR results of this 5% doped, 1 μm thick Co:ZnO sample, the

Co:ZnO - SAMPLES OF HIGH STRUCTURAL QUALITY

difficulties of finding magnetic resonance signals in the 10% Co doped ZnO samples discussed before become obvious: The layer thickness of the samples (100 nm) was optimized for synchrotron measurements, to prevent selfabsorption effects. With regard to the relative small peak to peak amplitude of the ESR signal of the 5% sample, a reduction of the absorbancy to one tenth results in very weak signals close to the detection limit. Further, the larger concentration of 10% causes a pronounced line broadening. The line broadening of the absorption curve itself is expected to scale with the concentration p as $\sqrt{p}$ [69]. However, a change of the FWHM Γ of the absorption curve leads to a reduction of the peak to peak amplitude of the first derivative as $1/\Gamma^2$ [57].

In conclusion the absence of any detectable ESR signals related to Co in 10% doped samples of high structural quality is attributed to line broadening effects, as experimentally observed in powders of Co:ZnO nanocrystals and a 5% doped Co:ZnO RMS grown sample.

5.2.2 Co^+-ion implantation in ZnO - low dopant concentration

In view of results and reports on highly dilute Gd:GaN (10^{-16} Gd/cm^3 [125]) low dopant concentrations were also realized in Co:ZnO by FIB implantation. Concentrations from $1 \cdot 10^{18}$-$1 \cdot 10^{20}$ Co/cm^3 were chosen. A typical implantation area extends over 4×4 mm. Co^+-ions were implanted with an energy of 100 keV. The penetration profile as calculated by SRIM [81] resulted in layer thicknesses of 100 nm.

The high structural quality as measured by XRD of the chemically grown ZnO substrates was not affected by the implantation - even for the highest implantation doses.

Magnetic properties of all samples have been characterized by SQUID magnetometry. Analogous to high dopant concentrations a paramagnetic behavior could be observed. Ferromagnetic-like signatures were absent or below the detection limit as discussed in 2.3. Results are reported in [149].

5.3 Co/CoO nanoparticles

The limit of phase segregation in Co:ZnO can be considered to be metallic Co or partly oxidized Co nanoparticles, which are so densely packed in the material that dipolar interaction of (residual) magnetic moments sets in. Even though investigations on Co/CoO nanoparticles were not the primary goal of this work, results presented in this section will be occasionally referred to in following sections. Fully oxidized clusters of CoO are expected to show only a reduced magnetic signal, since most magnetic moments are compensated by the antiferromagnetic coupling. The cobalt particles investigated in this section were exposed to ambient conditions, therefore oxidation processes are very likely. Resulting core-shell structures of oxide and metal are reported for various compounds in literature [150].

Results of a Co nanoparticle population with a mean diameter of 13.75 nm (stan-

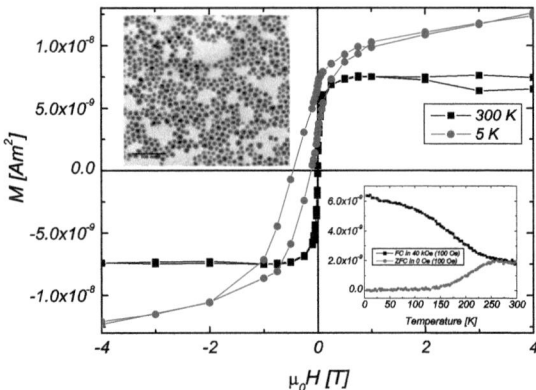

Figure 5.8: Magnetic characterization of Co nanoparticles with a mean diameter of 14 nm by SQUID. The shifted hysteresis at low temperatures represents an exchange bias effect due to the antiferromagnetic CoO outer shell of the nanoparticles. The step like behavior without coercivity at T = 300 K indicates superparamagnetic particles. In the inset the FC/ZFC curves are shown. Blocking temperatures around 260 K can be estimated.

dard deviation 2.85 nm (± 20%) [8]) are exemplarily shown in figure 5.8. The particles were spin coated on a sapphire substrate and afterwards subject to magnetization measurements. The hysteresis curve at T = 5 K reveals a large coercivity and a clear shift on the field-axis. This exchange bias effect is well known and was first reported in 1956 for Co/CoO [151]. In turn, the occurrence of a field shifted hysteresis indicates the partly oxidation of the particles. Note that an exchange bias effect crucially depends on the cooling conditions. If the particle is cooled through the Neel temperature of the antiferromagnetic CoO (293 K) with an applied field a preferential pinning of the magnetic moments of the ferromagnet in the field direction will occur, resulting in an exchange bias. In case of zero field cooling no exchange bias is expected.

The hysteresis taken a 300 K consistently shows no such shift. The curve exhibits an S-shape like magnetization behavior of the particles without any coercivity. This is expected for superparamagnetic particles above their blocking temperature.

The inset of figure 5.8 shows the FC/ZFC of the particles. Consistently the two curves merge around 260 K indicating a blocking temperature of the particles around this value. Due to the relatively broad size distribution of the particle population (standard deviation = 2.85 nm (± 20%)) individual blocking temperatures will also spread over a range of temperatures.

In the upper part of figure 5.9 a) the magnetic resonance spectrum taken at T = 5 K and $\theta = 15°$ is shown. The central feature around 330 mT consists of the cavity and cryostat signal and the superimposed paramagnetic resonance of Mo^{3+} embedded in the sapphire substrate matrix (see section 3.3). The broad feature around 150 mT is also known from pure sapphire.

The lower part of figure a) presents a color code plot of angular scans taken with 10 degree steps. The assignment of the resonance signals to the colors can be found in the upper part of the figure on the right side. Most features in the color code plot are straight vertical lines proving no angular dependence. The only clearly shifting feature is the narrow line of the Mo^{3+} impurity in the substrate.

The lack of a resonance signal of the Co/CoO nanoparticles can be understood if one assumes blocked particles, with arbitrary fixed magnetizations due to their individual anisotropy. Figure 5.9 b) shows a similar plot at an elevated temperature of T = 260 K. In the single spectrum in the upper part of the plot a broad, nei-

---

[8]statistical figures are derived from 579 analyzed particles

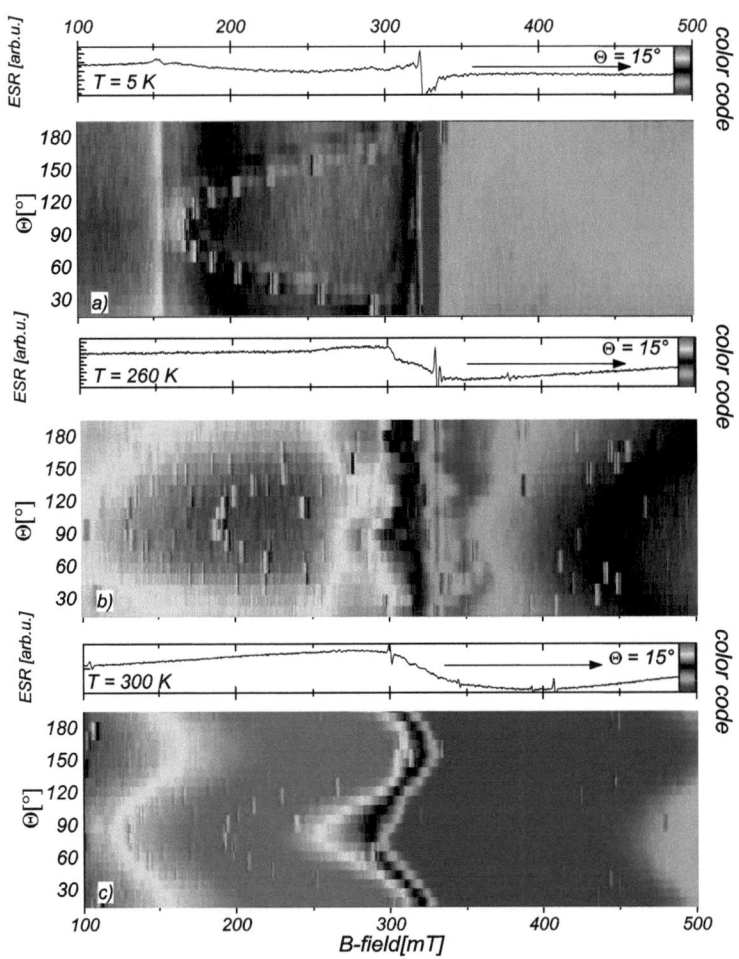

Figure 5.9: Magnetic resonance signals of Co/CoO nanoparticles and their angular dependence at temperatures of 5 K, 260 K, and 300 K. In the color plots the amplitude of the ESR signal is assigned to a respective color. In a) the only angular dependent signal originates from an impurity in the sapphire substrate (Mo^{3+}). b) Fractions of unblocked particles are likely to cause the onset of angular dependency. c) unblocked particles showing uniaxial anisotropy due to dipolar interactions.

Co/CoO NANOPARTICLES   115

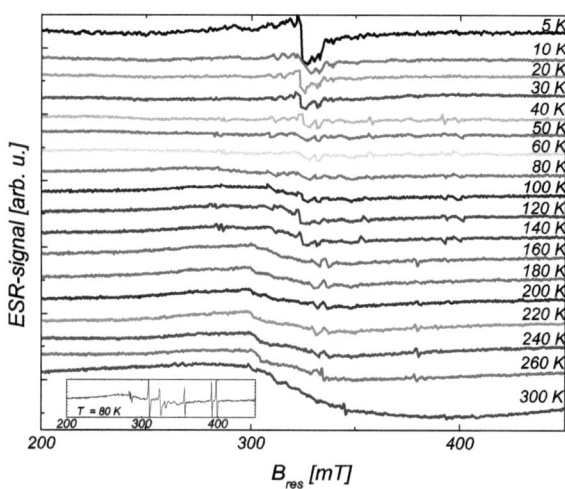

Figure 5.10: Temperature dependence of magnetic resonance spectra of Co nanoparticles. For better visibility data points of paramagnetic centers of the sapphire substrate have been removed. The inset shows exemplarily the complete 80 K spectrum. All spectra were taken in out-of-plane geometry ($\theta = 0°$). Only the 300 K data show a clear angular dependence as shown in figure 5.9 c). The central feature at T = 5 K is due to the cryostat.

ther Gaussian nor Lorentzian shaped, feature is visible at about 310 mT (roughly g = 2.16, Co bulk [150]). Such signatures are often reported in literature for superparamagnetic ensembles of nanoparticles [152]. Note that the superimposed narrow paramagnetic resonance spectrum originates not from Mo^{3+} but from Cr^{3+}-ions, also present in the sapphire substrate (see section 3.3). In the lower part of figure 5.9 b), the color code plot, slight angular dependent deviations of the central feature are apparent. This may originate from a small fraction of interacting particles, which is consistent with the blocking temperature range found by SQUID.

Finally figure 5.9 c) shows the magnetic resonance signal of the ensemble of nanoparticles at T = 300 K, above the Néel temperature of CoO (T_N = 293 K). Already in the upper spectrum for $\theta = 15°$ an increased intensity of the broad central feature is obvious. This indicates an enhanced amount of nanoparticles contributing to the signal. The color code plot unambiguously shows the uniaxial angular dependency of the resonance signal. The lower resonance field may be attributed to an easy axis in the plane and vice versa the high resonance field to an out-of-plane hard axis. A similar behavior is found for different ensembles of magnetic nanoparticles and has been reported in literature [152].

Figure 5.10 shows the temperature dependent development of the magnetic resonance signals of the Co/CoO nanoparticles. From roughly 80 K on a broad feature starts to appear around 310 mT in the spectra indicating unblocked, superparamagnetic particles. The onset of unblocking of the supermoments at 80 K does not agree with the blocking behavior found by SQUID. Aging effects are most likely to cause this discrepancy: SQUID measurements were performed about one month before ESR measurements. During that time the particles were exposed to ambient condition probably causing further oxidation of the particles.

Note that the central broad feature present at T = 5 K is due to the cryostat and Mo^{3+}. Its intensity weakens with increasing temperature and hardly perturbs the signal of the nanoparticles at elevated temperatures.

5.4 Clustering in Co:ZnO

The theoretical claim of ferromagnetic Co:ZnO was based on the premise of a homogeneous doped material of perfect structural quality. Experimentally in the case of the DMS Co:ZnO "several pitfalls" [153] have to be avoided, since ferromagnetic signatures might be easily of extrinsic origin. The problem of contamination and the limitations of SQUID measurements has been discussed in section 2.3. This section will address the problem of phase separation and clustering. Besides precipitates of ferromagnetic cobalt also ZnCo [143] alloys and all kinds of uncompensated magnetic moments of CoO nanoclusters have to be taken into account.

5.4.1 Clustering in Co:ZnO grown under oxygen deficiency

Some reports on ferromagnetic Co:ZnO claim oxygen vacancies to be crucial for the development of a long range magnetic order [154, 155]. The following mainly refers

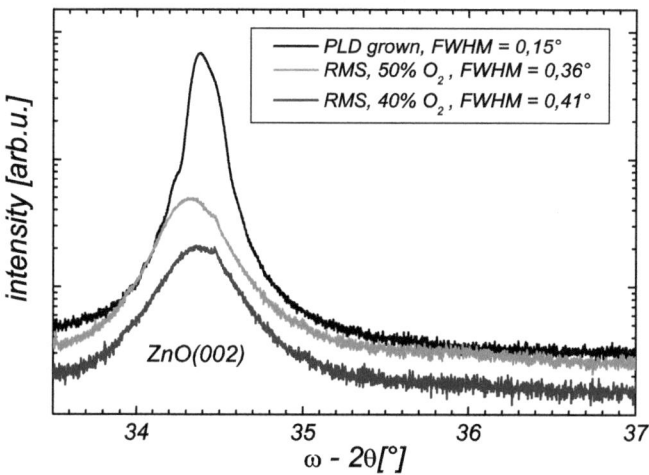

Figure 5.11: XRD-spectra of 10% doped Co:ZnO film grown by RMS under oxygen deficiency compared with best quality PLD growth. Reduced structural quality is visible.

to a series of samples grown by RMS under reduced oxygen partial pressure. The ratio of the sputter gas Ar to the reactant O_2 was changed from 10:1 to 10:0.4. Figure 5.11 shows XRD 2θ-scans of the ZnO(002) reflection of highest structural quality and reduced structural quality due to oxygen deficient growth. All samples show a clear peak close to the position of bulk ZnO of 34.4°. The FWHM of 0.15 of the PLD grown sample (SC042807B) is plotted as reference for the best quality. The sputtered samples grown at a reduced oxygen partial pressure of 40% (50%) - with respect to the optimized oxygen content - reveal a FWHM of 0.41° (0.36°) which indicates an increasing amount of structural imperfections. The peak intensities cannot be directly compared, since sample areas and thicknesses are different. Note that XRD yields no indications for secondary phases. However, recently faint traces of clustering were found by XRD in films grown under comparable reduced oxygen pressure conditions and/or elevated temperatures [156].

Figure 5.12 compiles the transition from paramagnetic to superparamagnetic be-

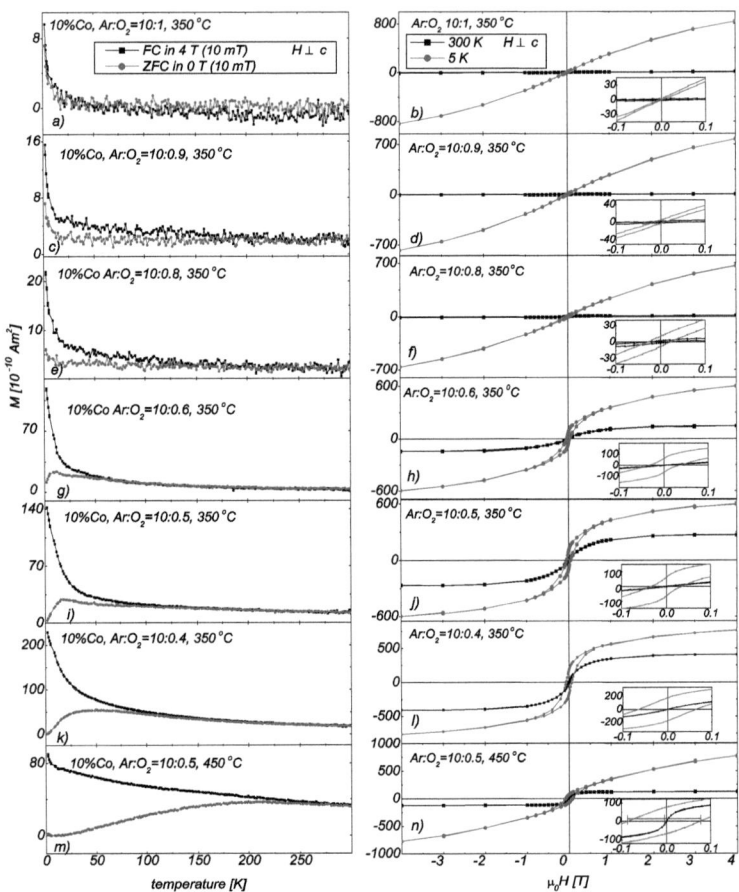

Figure 5.12: Development of magnetic properties with reduction of O_2 partial pressure while RMS growth as measured by SQUID. The left column shows the increase of the splitting of the FC/ZFC curves. For reduced O_2 a maximum in the ZFC measurement can be observed, which is indicative for clustering. The right column presents hysteresis curves which show increasing coercivity with reduced O_2 content. In the inset of n) even a small exchange bias effect is visible.

havior due to reduced oxygen partial pressure. The left column shows the FC/ZFC curves, which are all measured with an applied field of 10 mT, whereas in the right column M(H) plots are shown. The magnetization curves are plotted for temperatures of 5 K and 300 K, respectively. In the insets a close-up of the low field range of the hysteresis is shown.

The Ar:O_2 ratio which is known to yield the best structural properties at a preparation temperature of 350 °C is 10:1. In a) no FC/ZFC splitting is visible. The 1/T-like increase towards low temperatures is consistent with pure paramagnetic behavior in agreement with the s-shaped low temperature M(H) curve shown in b). A small FC/ZFC splitting and an opening of the hysteresis are shown in plots c) – f). In g) the ZFC curve starts to show a maximum at low temperatures. This temperature dependence is very characteristic for unblocking of supermoments, which subsequently start to behave superparamagnetic. The respective hysteresis at low temperature becomes more squarish and the M(H) curve at 300 K clearly deviates from zero and reaches saturation values at 3 T.

For even further reduced O_2 partial pressure this behavior is enhanced, which results in increased coercivity and remanence. The peak in the ZFC curve shifts to higher temperatures which indicate a higher blocking temperature of the respective cluster ensemble and indicates an increased cluster size.

The SQUID measurements of a sample grown with just 40% of the optimal O_2-pressure, are shown in figure 5.12 k) and l). Coercivity and remanence at low temperatures are indicative for a ferromagnetic-like behavior. The FC/ZFC exhibit a splitting up to approximately 70 K. The maximum in the ZFC curve reveals ferromagnetic clusters as possible origin. The shape of the hysteresis at T = 300 K is consistent with a superparamagnetic behavior as it is also observed by ESR measurements (figure 5.13) and will be discussed later on.

Note that in case of the last plots m) and n) the O_2 content was 10:0.5 and the growth temperature was increased to 450 °C. The maximum of the ZFC curve is broadened and the separation to the FC curve persists up to 250 K. This indicates a different cluster population with a wider distribution of blocking temperatures which are on average much enhanced compared to i). In n) the room temperature hysteresis shows a saturatation at the lowest fields (< 1 T) compared to all the other samples, revealing larger supermoments. From the coercive fields of the inset of n) even a faint exchange bias effect can be seen. This points towards small

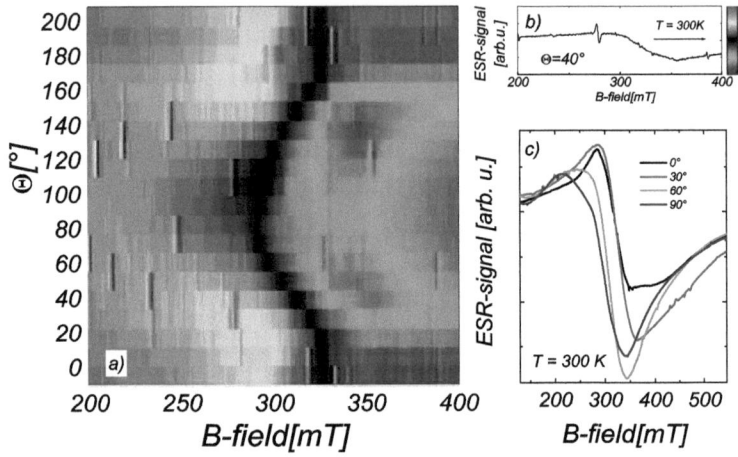

Figure 5.13: Magnetic resonance signal of the sample grown under lowest oxygen pressure. a) Angular dependence of the resonance field depicted in a color code plot. The color black corresponds roughly to the center of the line. b) A broad magnetic resonance observable at room temperature exemplarily shown for $\theta = 40°$. The small bar to the right shows the color code used for the color plot. c) Angular dependence of the line shape. Interacting superparamagnetic clusters are likely to cause the signal. The narrow resonances visible in a) and b) are due to paramagnetic impurities of the sapphire substrate.

fractions of CoO as additional phase in the sample, since this phenomena is known from Co/CoO nanoparticles (see section 5.3).
In summary, figure 5.13 shows the transition of from a purely paramagnetic behavior of Co:ZnO to a ferromagnetic-like/superparamagnetic one by reduction of the oxygen partial pressure during growth. Low oxygen pressure and high growth temperatures presumably favor the growth of metallic Co cluster, thus resulting in larger supermoments and higher blocking temperatures.
Figure 5.13 a) - c) shows the magnetic resonance signals observed at room temperature and X-band frequency of the sample grown under the lowest oxygen partial pressure (plot k) and l) of figure 5.12). In b) a broad resonance with a peak to peak

CLUSTERING IN Co:ZnO

width of approximately 50 mT is visible. Figure 5.13 a) shows the angular dependence using a color code related to the amplitude of resonance curve, exemplarily shown on the right side in b). The color plot reveals a uniaxial behavior of the resonance field. The resonance field shifts from 340 mT to 300 mT which is visible in both figures 5.13 a) and c). This anisotropy is likely to originate from dipolar interactions of the unblocked supermoments of the clusters. The line shape cannot be well fitted neither by Lorentzian nor Gaussian, especially at angles different from $\theta = 40°$. The change of the line shape is shown exemplarily for four angles in c). All spectra in c) have been smoothed to remove the narrow paramagnetic impurity signals. The line shape changes from triangular-like (30°) to more rounded in the in-plane case. While the 30° spectrum appears symmetric to the resonance field, the 90° scan is asymmetric. This behavior is indicative of an inhomogeneously broadened line, consisting of several contributions of an ensemble of superparamagnetic clusters.

In figure 5.14 the appearance of the cluster related central resonance signal of a 10% Co doped ZnO film (080917) with increasing temperature is shown. The onset of the signal becomes visible after subtraction of a background signal. The spectrum taken at T = 60 K was used for this purpose. Similar to Co/CoO nanoparticles (figure 5.10) no angular dependent resonance signal stemming from the Co:ZnO film was found at low temperatures. Note that in ESR measurements higher blocking temperatures of nanoparticles compared to SQUID measurements are expected due to the Arrhenius law (see e.g. [157]).

Figure 5.15 models the superposition of several Gaussian resonance signals to get a qualitative understanding of the observed line shapes. A normalized Gaussian absorption curve is described by:

$$y = y_{max} \exp\left[\frac{-4\ln(2)(B - B_{res})^2}{\Gamma^2}\right] \quad (5.2)$$

With Γ being the FWHM and $y_{max} = \frac{1}{\Gamma}\sqrt{\frac{\ln(2)}{\pi}}$. The first derivative with respect to the field is then:

$$\frac{dy}{dB} = -y_{max}\frac{8\ln(2)(B - B_{res})}{\Gamma^2} \exp\left[\frac{-4\ln(2)(B - B_{res})^2}{\Gamma^2}\right] \quad (5.3)$$

The resonance fields in figure 5.15 are symmetrically distributed around a central value (g = 2 ("$\frac{\omega}{\gamma}$")). The weighting of the individual signals is chosen to follow roughly a Gaussian (black symbols). This implies smaller anisotropies to be more

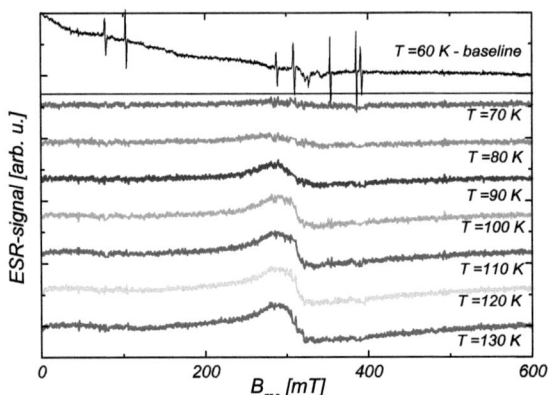

Figure 5.14: Temperature dependent appearance of cluster related resonance signal in 10% Co doped ZnO sample grown at reduced O_2 partial pressure (Ar:O_2 10:0.5). The development of the typical broad resonance around g = 2 ("$\frac{\omega}{\gamma}$") becomes clearly visible after subtraction of the T = 60 °C spectrum and removal of paramagnetic impurity signals.

likely than larger ones. The resulting sum (bold red line) is lower in peak to peak intensity than the central individual signal components. The latter reflects the partly destructive superposition of the signals.

Thus the absorption I(B) in case of infinite resonance contributions yields:

$$I(B) = \int_{-\infty}^{\infty} y_{max}(B_{res}) \exp\left[\frac{-4\ln(2)(B-B_{res})^2}{\Gamma^2}\right] dB_{res} \tag{5.4}$$

and the derivative is given as:

$$\frac{dI(B)}{dB} = \int_{-\infty}^{\infty} -y_{max}(B_{res}) \frac{8\ln(2)(B-B_{res})}{\Gamma^2} \exp\left[\frac{-4\ln(2)(B-B_{res})^2}{\Gamma^2}\right] dB_{res} \tag{5.5}$$

Equation 5.5 represents the convolution of the first derivative of a Gaussian with the function describing the weight of the resonance components ($y_{max}(B_{res})$). The peak to peak amplitude of $\frac{dI(B)}{dB}$ crucially depends on $\frac{dy_{max}(B_{res})}{dB_{res}}$, since $\int_{-\infty}^{\infty} \frac{dy}{dB} = 0$. This means that the detectability of a signal is correlated to the anisotropy distri-

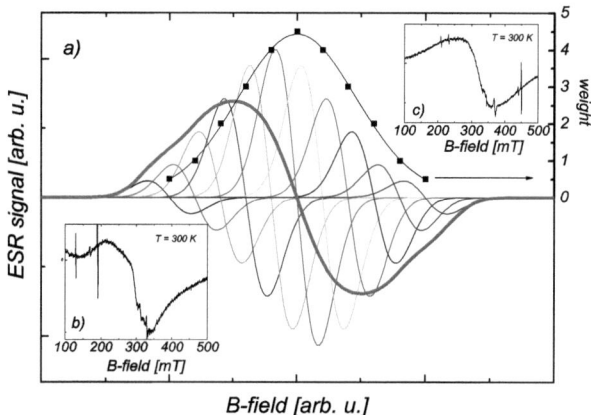

Figure 5.15: a) Superposition of eleven 1st derivatives Gaussian lines. The red bold line is the sum of the curves. The black symbols show the chosen weight distribution following roughly a Gaussian function. The two insets are measured signals of b) Co:ZnO sample grown under O_2 deficiency and c) Co nanoparticles.

bution of clusters. A similar amount of clustered material may be visible in ESR measurements in case of $|\frac{dy_{max}(B_{res})}{dB_{res}}| \gg 0$, whereas the signal might vanish for a broad distribution of resonance fields, i.e. $|\frac{dy_{max}(B_{res})}{dB_{res}}| \rightarrow 0$.

However, the assumption of crystalline anisotropies of the individual cluster of nanoclusters is in contrast to the high temperatures where the uniaxial angular behavior is observed experimentally (see figure 5.15 b) and c)). As shown in figure 5.12 typical blocking temperatures are far below room temperature. Therefore, the angular dependence of shape and resonance field of the broad resonance signals is more likely to originate from dipolar interactions of the individual particles. These dipolar interactions result in an additional anisotropy - a shape anisotropy of the particle ensemble. Each particle can have a different angular dependence, since the internal fields are likely to be inhomogeneous. This in turn can lead to angular dependent changes of the superposition, as observed experimentally. The two insets of plot 5.15 present measured data at T = 300 K on b) a Co:ZnO sample grown under O_2 deficiency and c) Co/CoO nanoparticles which were deposited on sapphire.

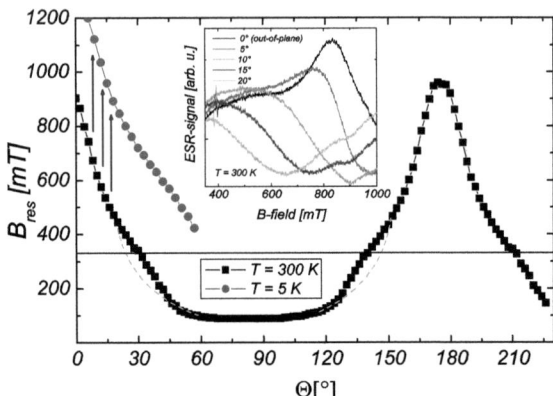

Figure 5.16: Angular shift of ferromagnetic-like feature in ZOC5164. The inset shows exemplarily the change of the very distorted signals at angles up to 20°.

These considerations have to remain phenomenological, since a modeling of experimental data would require detailed knowledge about cluster size distribution, individual anisotropies, size of magnetic moments and spatial distribution of the supermoments. However, a qualitative understanding of the origin of the observed line shapes can be obtained. The presented results and the interpretation as superparamagnetic ensemble are in agreement with recent results of other groups [158].

5.4.2 Metallic precipitations in Co:ZnO

Bulk cobalt is known to exhibit a large magnetic anisotropy. Magnetic resonance spectra presented so far have only shown shifts of the resonance field of about 200 mT. These spectra are likely to originate from metallic clusters in a superparamagnetic state, i.e. $T > T_B$, as mentioned before.

Figure 5.16 shows the magnetic resonance fields measured on a 5% Co doped sample grown on r-plane sapphire by PLD (ZOC5164). The inset of figure 5.16 exemplarily plots the resonance curves at T = 300 K for angles form 0° to 20°. The signal strongly changes its shape with angle which significantly complicates the analysis of the data. As a consequence the given values of the resonance field have large

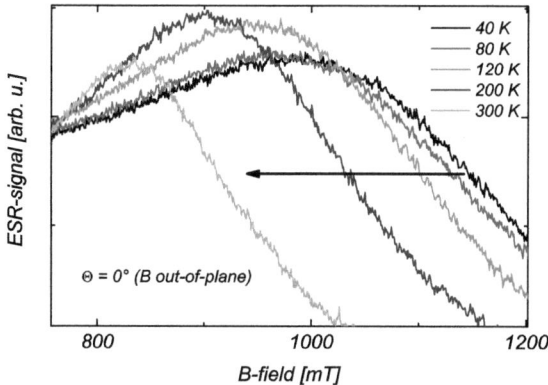

Figure 5.17: Temperature dependent shift of ferromagnetic-like resonance. The decrease of the anisotropy is typical for a ferromagnetic signal.

uncertainties. A kink in the angular dependence of the resonance field around 400 mT is visible and might stem from an additional signal. At lower fields the gray dotted lines are guide to the eye for what could be expected from a resonance of a homogeneous system. As discussed in the previous sections the line shape is indicative of a superposition of several components.

Figure 5.16 reveals some remarkable features: i) Resonance fields of up to 1 T in out-of-plane geometry at room temperature with a shift below 0.1 T for in-plane geometry proving a very large anisotropy. ii) In contrast to previously discussed samples, magnetic resonance signals are also present at very low temperatures. At T = 5 K resonance fields are shifted to higher values, revealing an increased anisotropy (blue arrows). This behavior is shown in more detail in figure 5.17. In the out-of-plane position, measured from 40 K to 300 K, the resonance field changes about 200 mT towards the isotropic g-value ("$\frac{\omega}{\gamma}$"). Such a shift is very typical for ferromagnetic resonance. It reflects the competition between anisotropy and thermal energy.

The ferromagnetic response up to room temperature of this sample (ZOC5164) as it is found by magnetic resonance measurements is also corroborated by SQUID measurements which are shown in figure 5.18. Coercivity and remanence at low fields indicate a ferromagnetic Co fraction. The curvature of the low temperature hysteresis at high fields suggests additional paramagnetic contributions, possibly by properly incorporated Co^{2+}-ions on Zn^{2+} sites. At high temperatures the paramagnetism is expected to be negligible. Conclusively the high temperature M(H) curve shows a steplike behavior with clear remanence and coercivity (upper inset) typical of a ferromagnet. The latter is also corroborated by the FC/ZFC measurements, shown in the lower inset of figure 5.18. The separation of the two curves at T = 300 K is only reduced to 85% of the T = 5 K value, thus suggesting a high T_C. The absence of a maximum in the ZFC curve is consistent with a low temperature magnetic resonance signal, thus corroborating no blocking behavior.

Further clarification of the properties of the sample could be achieved by synchrotron measurements. Figure 5.19 presents XANES and XLD spectra of the K-edge of Zn and Co, respectively. The measurements were performed in grazing incidence geometry in fluorescence yield as described in section 2.4.

In Figure 5.19 a) the XANES at the Zn K-edge are shown. For comparison the respective spectrum of a Co:ZnO sample (080428) grown by RMS under optimized conditions is plotted additionally (dotted lines). All spectral features of this spectrum can be found in the XANES of sample ZOC5164, indicative of well incorporated, tetrahedrally coordinated Zn^{2+}. The XLD shows the typical wurtzite signature. However, the amplitude of the XLD signature of sample ZOC5164 at the Zn K-edge is reduced to 84% of the referece sample 080428, revealing less structural quality.

Figure 5.19 b) presents the respective plots for the Co K-edge. The absorption spectra differ significantly with respect to the spectrum of the sample 080428. The fine structure is much less pronounced and individual peaks are less distinct. The minimum between the preedge feature and the main peak, visible in the spectrum of the best quality sample (080428), is filled. The latter is considered as strong indication for metallic cobalt. The XLD signal is much more reduced compared to the Zn K-edge. A wurtzite pattern is still present but in this case the amplitude is reduced to 51% of the XLD amplitude of the sample grown under optimized condition. Therefore the structural analysis by XLD indicates that only half of the Co atoms in this sample are on Zn substitutional sites in proper wurtzite environment. Figure 5.20 shows the generation of a XANES spectrum by a weighted superposition (red) of two XANES spectra. The spectrum of sample 080428 (black), which is one

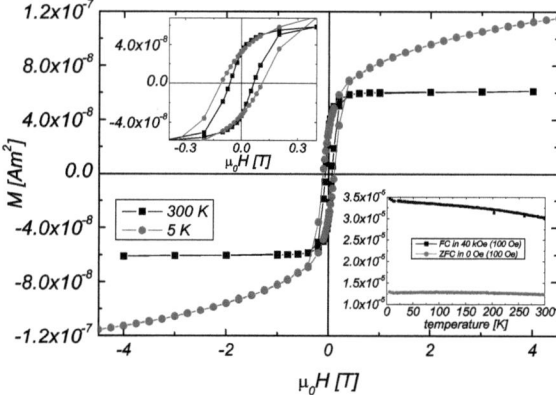

Figure 5.18: SQUID measurements on sample ZOC5164. The steplike behavior, remanence and coercivity at T = 300 K corroborates ferromagnetic-like behavior. The s-shape low temperature hysteresis indicates additional paramagnetic contributions. The FC/ZFC measurements as shown in the lower inset suggest a T_C far above room temperature.

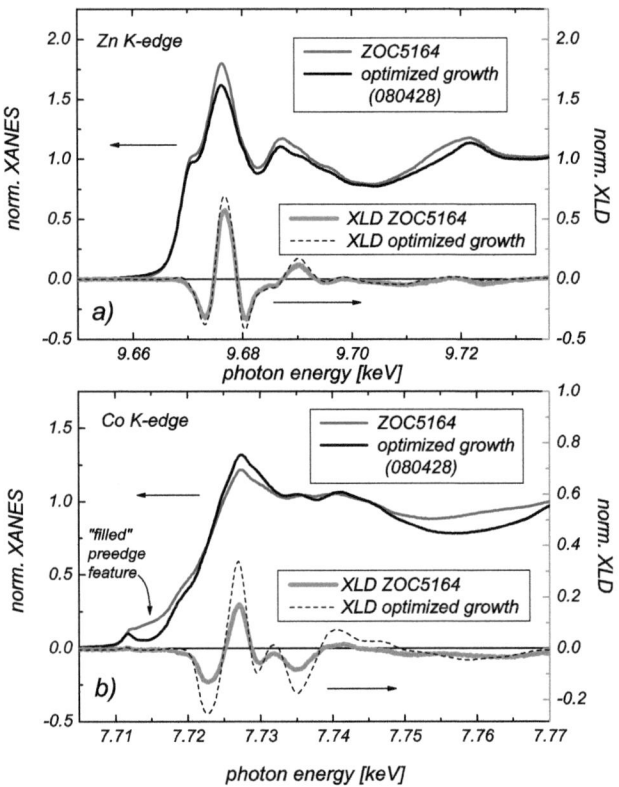

Figure 5.19: XLD analysis of sample ZOC5164. a) The absorption edge and the XLD of Zn indicate reduced structural quality but reveal still clear wurtzite structure. b) At the Co K-edge strong deviations from best quality samples are visible. The XLD indicates that less than 51% of Co atoms are in proper wurtzite environment. The "filled preedge" feature is typical for metallic cobalt.

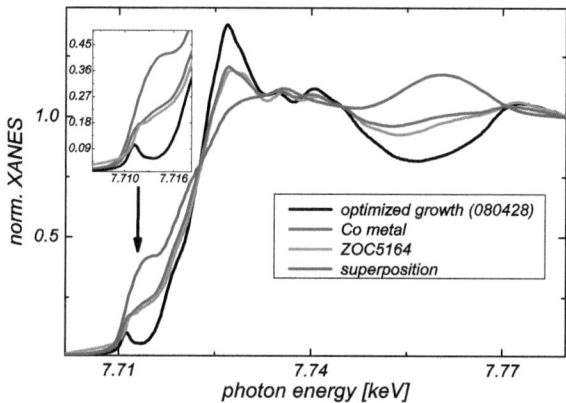

Figure 5.20: Generation of XANES spectrum of ZOC5164 by weighted superposition of a Co metal spectrum and a spectrum of a sample which was grown under optimized conditions. The weight of each spectrum is determined by the XLD amplitude.

of the structural best samples as measured consistently by XRD and XLD, is used as a reference spectrum for substitutional incorporated Co^{2+}. Its weight for the superposition is given by the XLD amplitude of sample ZOC5164 relative to the XLD amplitude of sample 080428. For the remaining fraction (49%) of the superposition the XANES spectrum of hcp Co metal (purple) is chosen.

The experimental XANES of sample ZOC5164 (green) in figure 5.20 agrees remarkably well with the weighted superposition. In particular the filling of preedge feature, which is a measure for the metallicity of a sample, is reproduced well(inset).

Finally the element specific magnetic properties were investigated by XMCD as shown in figure 5.21. The inset shows the respective dichroism and the absorption edge at T = 6.5 K. The dichroic signal is dominated by two triangular-like shapes, below and above the zero line. Superimposed one can find the dichroic peaks as known from samples grown by RMS under optimized conditions and shown in figure 5.3. The triangular-like signature at 7718 eV (green arrow) is known to be indicative for Co in its metallic state The main part of figure 5.21 presents M(H) curves that were taken at two different energy values. The green M(H) curve was taken at the maximum of the spectral weight of the metallic Co XMCD (green arrow) whereas the blue curve was taken at the the preedge feature (blue arrow). The M(H) curve taken at an energy of 7718 eV shows a steplike switching of the magnetization close to (not at) zero field and is saturated at about 0.5 T. This is expected for cobalt in the metallic state. Observing the magnetic behavior at the preedge at an energy of 7711 eV in contrast shows a s-like shape, as expected for paramagnetic Co^{2+}. No saturation is visible even at fields of 6 T. Nevertheless the steplike shape at low fields is also present. This is due to the extent of the spectral weight of the metallic cobalt phase which underlays the paramagnetic one.
[159].

Therefore consistently SQUID, ESR and XMCD show a metallic Co phase in this sample up to room temperature. In contrast to probably clustered Co:ZnO samples grown by RMS the metallic Co phase of sample ZOC5164 is likely to originate from larger Co precipitates. Instead of a superparamagnetic behavior, SQUID hysteresis and magnetic resonance signals are observed from T = 5 K up to room temperature. The metallic Co phase is likely to originate from the different growth conditions: The sample was grown by PLD on r-plane sapphire ($01\bar{1}2$) at temperatures up to

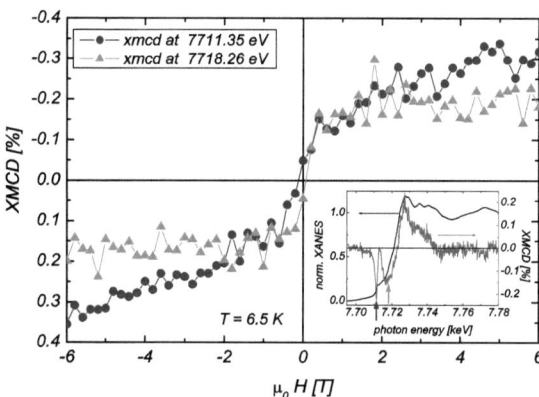

Figure 5.21: XMCD analysis of ZOC5164. M(H) curves taken at two different energies show different weight of the para-/ferromagnetic contribution. Superimposed on the known XMCD signal of Co^{2+} one can identify a triangular shape which is typical for metallic Co (inset).

600 °C. The crystallographic orientation [9] of the sapphire substrate possibly favors metallic precipitations in addition.

5.5 Annealing effects

Several annealing experiments have been conducted to examine possible effects of the structure, carrier- or defect concentration, on the magnetism of Co:ZnO. In particular high vacuum (HV) annealing experiments and annealing under O_2 atmosphere will be treated in this section. HV-annealing has frequently been reported to enhance the ferromagnetic-like behavior in Co:ZnO due to structural improvement, but the problems of metallic cobalt precipitation has been reported already in 2003 [160]. O_2-annealing in turn has been reported to decrease the ferromagnetic properties. This is in accordance with the model of magnetic polarons suggested for DMS [161]. The respective coupling between the magnetized spheres is expected to

---
[9]all other Co:ZnO samples discussed in this work were grown on c-plane sapphire (0001)

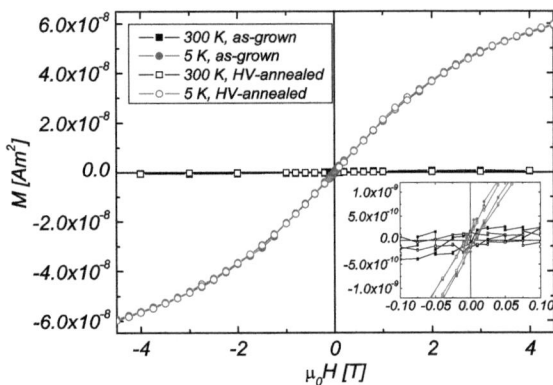

Figure 5.22: No annealing effect concerning magnetization behavior in best grown Co:ZnO (080428). The respective M(H) curves before and after the annealing procedure are almost identical. The inset enlarges the low field behavior of the respective hysteresis curves. Residual remanence and coercivity are below the artifact level as explained in section 2.3 and [87].

be enhanced by oxygen vacancies [154, 162].

5.5.1 High vacuum annealing of best quality samples

This first part will deal with samples which are considered to be of high structural quality, i.e. a high percentage of Co^{2+}-ions substituting Zn^{2+}, as described in section 5.2. First, the magnetic properties after an annealing procedure under HV conditions will be described which will be later compared with the local structural properties as they were measured by XLD.

Figure 5.22 shows magnetization curves taken by a SQUID magnetometer for a sample grown under optimized conditions resulting in high structural quality. This sample (080428) was annealed for 30 minutes at 450 °C under high vacuum conditions ($p < 1 \cdot 10^{-5}$ mbar). The hysteresis curves taken before and after the annealing procedure are indistinguishable at T = 300 K and at T = 5 K. Also the close–up of the small field range, shown in the inset, confirms no enhanced splitting of the

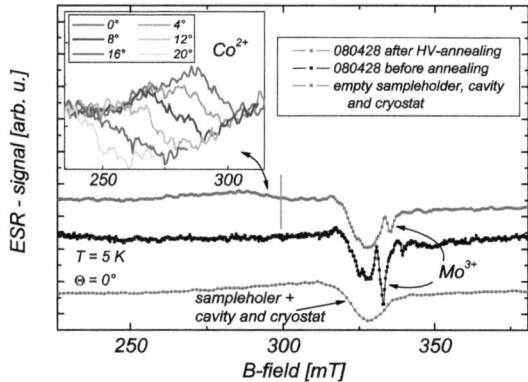

Figure 5.23: Single ion Co^{2+} ESR spectra observed after annealing. Note that the respective scans were taken with different spectrometer settings which cause for example the smoothing of the sample holder measurement (see text). The inset shows the shift of the small signal with sample angle. The Mo^{3+} spectrum has been discussed in section 3

M(H) curve. The remaining residual splitting is below what is considered to be the artifact level of SQUID measurements [87]. Therefore the paramagnetic behavior of the sample has not changed due to the annealing procedure.

In figure 5.23 magnetic resonance spectra at T = 5 K of the as-grown sample (black curve) and the annealed sample (red curve) are shown. In addition the signal of the empty sample holder at this temperature is plotted (gray curve). In contrast to the as-grown sample the spectrum of the HV annealed sample reveals a shallow broad resonance around 300 mT. The peak to peak width is roughly about 20 mT. The resonance field is consistent with the expected value for Co^{2+}-ions in wurtzite ZnO in $B \parallel c$ orientation. A yellow vertical line depicts the resonance field of the Co^{2+} single ion spectrum for out-of-plane geometry as discussed in section 3.2. In the inset of figure 5.23 the angular dependent shift of the weak signal is shown.

The ESR scan of the empty sample holder was measured with different spectrometer settings (less point density and bigger modulation amplitude) which results in a slightly smoothed spectrum. A dip, which is mainly caused by the quartz tube of the cryostat, is clearly visible. Even though this feature is in principle unwanted one can benefit from it as a reference concerning the signal sizes of the as-grown and annealed sample. The absence of the very weak Co^{2+} signal of the annealed sample in the as-grown sample becomes only credible by the comparison of the different spectra to the cavity signal.

The appearance of the Co^{2+} spectrum in the annealed sample is likely to result from a better structural quality caused by the annealing process. This means that more incorporated Co atoms are placed on substitutional sites, replacing Zinc and/or the crystal environment of the respective atoms has become better defined.

In summary of integral magnetic measurements, no changes of the magnetization behavior as measured by SQUID have been found, whereas in the ESR spectra of the annealed sample a weak magnetic resonance line of Co^{2+} becomes visible after annealing. However, this is consistent with the SQUID results, since this resonance signal is considered to reflect paramagnetism.

For further clarification the XLD signature was investigated. Figure 5.24 a) shows the respective XANES and XLD at the Zn K-edge. Obviously the changes in the absorption spectra are very faint. The resulting XLD shows a small improvement of about 3%. Note that the absolute XLD amplitude at the Zn K-edge of this magnetron sputtered sample (1.10) is comparable with that of best PLD grown samples (1.09, see figure 5.1). The annealing treatment obviously caused only a minor improvement of the local structure of the Zn cation environment.

At the Co K-edge - which is shown in figure 5.24 b) - the situation seems to be similar if one considers the absorption spectra only. Before and after annealing hardly any changes become visible. Especially the preedge feature remains unaffected. Slightly bigger changes can be derived from the XLD. Even though the absolute increase of the amplitude after annealing is only 0.04 the relative change is about more than twice (7%) as high as at the Zn K-edge, since the maximum XLD amplitude of the as-grown sample at the Co K-edge is 0.59. Therefore the annealing procedure enhances the relative amount of Co atoms in proper wurtzite environment. This was already suggested by the ESR measurements. However, this structural improvement does not alter the integral magnetic properties as measured by SQUID magnetometry.

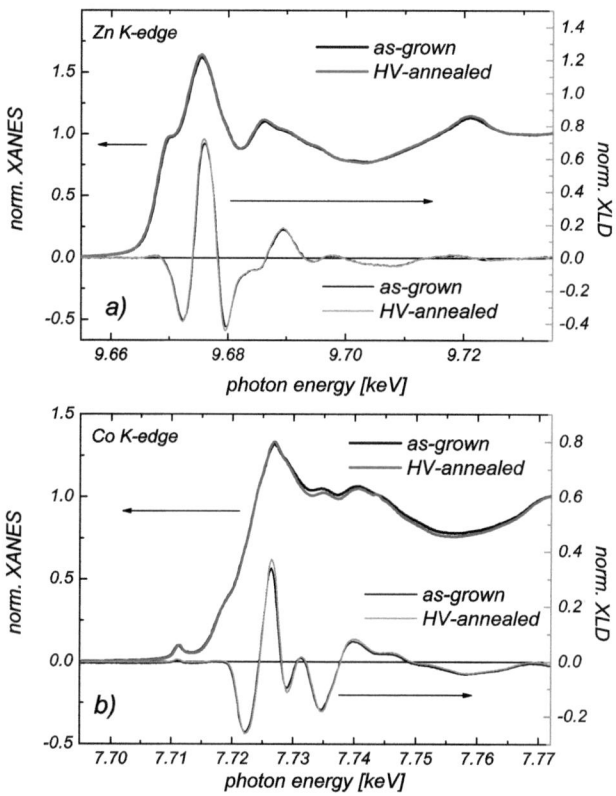

Figure 5.24: a) XANES and XLD at the Zn K-edge after HV-annealing of sample 080428. Hardly any changes are visible in the absorption spectra. The amplitude of the XLD signature shows a small structural improvement. b) Improvement of the XLD signal at the Co K-edge after HV-annealing. Even though the absorption spectra seem almost identical small deviations result in a by 7% increased XLD amplitude.

5.5.2 Annealing of samples with reduced oxygen content

In this section the annealing effects on samples which are grown under oxygen deficiency as introduced in section 5.4 will be presented. Note that most of the samples still have to be considered to be Co:ZnO of high structural quality considering XRD results [156].

Two contrary effects depending on whether the annealing procedure took place under HV conditions or O_2 flow will be discussed by their magnetic signature (SQUID) and correlations to the local crystallographic structure (XLD). At the end of this section the findings will be used to understand the respective ESR spectra.

The 10% Co doped sample 080312 was grown with a Ar:O_2 ratio of 10:0.5 and two parts of this sample were subject to HV and O_2-annealing procedures, respectively. This section mainly focuses on this sample, since it was also subject to thorough synchrotron measurements.

In Figure 5.25 SQUID measurements before and after HV-annealing of sample 080312 are compiled. In a) the respective hysteresis curves in the low field range are shown. After annealing coercivities and remanences are enhanced at both temperatures $T = 5$ K and $T = 300$ K, respectively. The high field range is shown in b), revealing an increased saturation magnetization at $T = 300$ K. In contrast, the $T = 5$ K data show a reduced magnetization of the annealed sample at a field of 4 T. This is probably caused by a reduced amount of paramagnetic defects after annealing. The respective FC/ZFC curves are shown in c). They are much more separated and no merging of the curves can be observed in the temperature range up to 300 K for the annealed sample. A maximum of the ZFC curve is not visible any longer in this temperature range; probably it is shifted above $T = 300$ K. Note that at temperatures below 10 K the increase of the magnetization of the annealed sample due to paramagnetic contributions is reduced (see e.g. figure 3.1).

Assuming the ferromagnetic-like response of the sample 080312a(HV) originates mainly from two contributions, a superparamagnetic and a paramagnetic one, the respective atomic fractions can be estimated:

At 5 K and a field of 4 T the magnetic moment is $5.92 \cdot 10^{-8}$ Am^2, which is 3.4 times the room temperature (300 K) value of $1.75 \cdot 10^{-8}$ Am^2. Assuming a Brillouin like behavior of the paramagnetic Co^{2+}-ions more than 80% of the saturation magnetization of the paramagnetic Co atoms should contribute at 5 K and 4 T. In contrast, the contribution at 300 K and 4 T can be neglected compared to a saturated super-

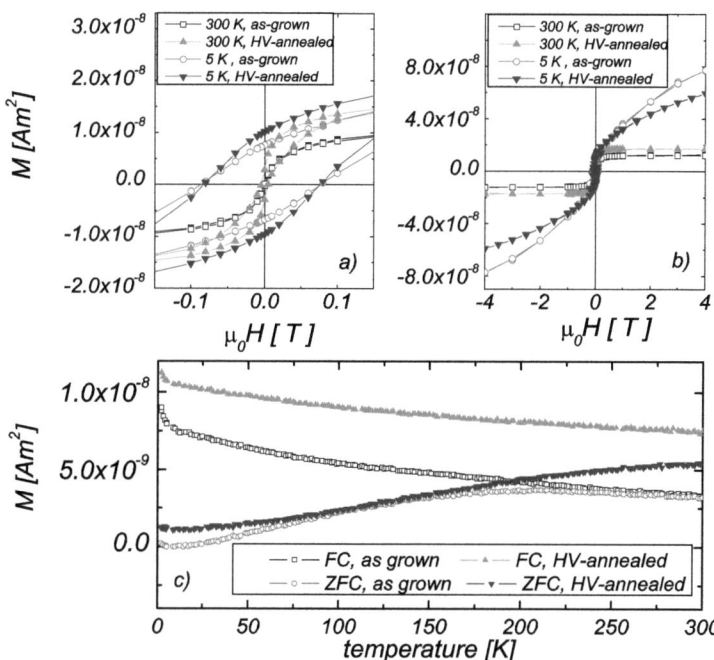

Figure 5.25: HV-annealing effect on magnetic behavior of sample (080312): a) Increased hysteretic behavior, b) high field data, c) FC/ZFC curves before annealing show indication for clustering. The unblocking of supermoments probably causes the maximum around 230 K. After annealing the separation of the two curves maintains up to room temperature indicating a blocking above T = 300 K. FC/ZFC curves were measured with an applied field of 10 mT.

paramagnetic contribution. From this the metallic fraction can be calculated:

$$\underbrace{\frac{\overbrace{5.92\ Am^2 - 1.75\ Am^2}^{param.frac.}}{0.8}}_{BF(5K,4T)} \approx \overbrace{5.21\ Am^2}^{total\ param.\ frac.} \rightarrow \frac{1.75\ Am^2}{5.21\ Am^2} \approx 34\%\ \text{metallic Co} \quad (5.6)$$

This calculation neglects the different magnetic moments of paramagnetic Co^{2+}-ions of 4.1 μ_B and 1.7 μ_B for metallic Co [24, 143]. Therefore the magnetic moment of paramagnetic Co has to be divided by 4.1 μ_B and the metallic one by 1.7 μ_B to derive the atomic fractions. Taking this into account the calculation yields:

$$\frac{5.92\ Am^2 - 1.75\ Am^2}{0.8} \approx 5.21\ Am^2 \rightarrow \frac{1.75\ Am^2/1.7\ \mu_B}{5.21\ Am^2/4.1\ \mu_B} \approx 81\%\ \text{metallic Co} \quad (5.7)$$

This result is in stark contrast to fractions derived from a XANES analysis described later on.

Figure 5.26 compiles the magnetization data before and after annealing under O_2 flow of sample 071115 [10]. This 10% Co doped sample was grown with a Ar:O_2 ratio of 10:0.6. In figure 5.26 the development of the sample's magnetic behavior is obviously contrary to the case of HV-annealing. The FC/ZFC curves, which had been separated up to 60 K in the as-grown sample overlap after the annealing procedure. The collapse of the FC/ZFC separation is also corroborated by the hysteresis curves which are plotted in the inset. A former big coercivity and remanence of 40 mT and 6·10^{-6} Am^2, have vanished, respectively. After the oxygen exposition even the low temperature hysteresis shows less magnetization than the high temperature hysteresis of the as-grown sample. Oxidation effects on clusters as discussed in section 5.4 are very likely to cause the break down of ferromagnetic features of the samples.

Element specific insight into the effects of the annealing processes was yielded by XLD measurements at the K-edges of host cation and dopant: The linear dichroism taken at the Zn K-edge of sample 080312 is shown in figure 5.27 a). The respective XANES reveals hardly any differences. The XLD in turn exhibits changes of the local host cation environment depending on the annealing procedure. The HV-annealing results in a small reduction of the amplitude whereas the O_2-annealing clearly improves the structural quality. The latter is plausible, since the sample was grown under O_2 deficient condition. Note that even after the O_2-annealing

---

[10]O_2 annealing effects, as observed by SQUID of sample 071115 are more pronounced compared to sample 080312a(O_2) and therefore shown for clarity.

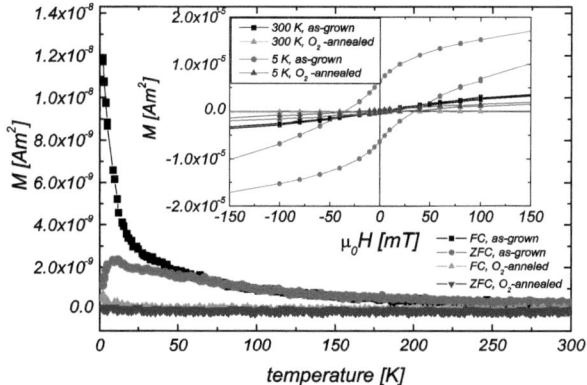

Figure 5.26: Effect on magnetic behavior of O_2-annealing on sample (071115). Former separated FC/ZFC curves match after annealing. The hysteresis curves (inset) corroborate the breakdown of any ferromagnetic like signature.

procedure the amplitude of the XLD is only 60% of the maximum XLD amplitude measured for samples of best structural quality (080428).

Figure 5.27 b) presents the XLD measurements on sample 080312 and its annealed versions at the Co K-edge. In view of the SQUID results the preedge feature is of special interest, since it indicates different degrees of hybridization between the 3d and 4p electron states. Note that in metallic cobalt no preedge feature is visible at all (see e.g. XANES spectra in figure 5.20). Considering the previous results on the Zn K-edge, figure 5.27 b) shows some remarkable results. The XANES for the as-grown and HV-annealed sample resemble very much - including the preedge feature, similar to what was already observed at the Zn K-edge. Contrary to this, in case of O_2-annealing a prominent second peak has evolved leading to a double peak structure very different from the other absorption spectra.

Clearly the most astonishing feature seen in figure 5.27 b) is the almost unchanged XLD signature - no matter which treatment the sample has experienced.

Obviously the structural changes of cation and dopant due to the annealing procedure do not affect the XLD signal of the Co. To understand this one has to recall what is visible by XLD and what is not. Well incorporated Co^{2+} on Zn lattice sites

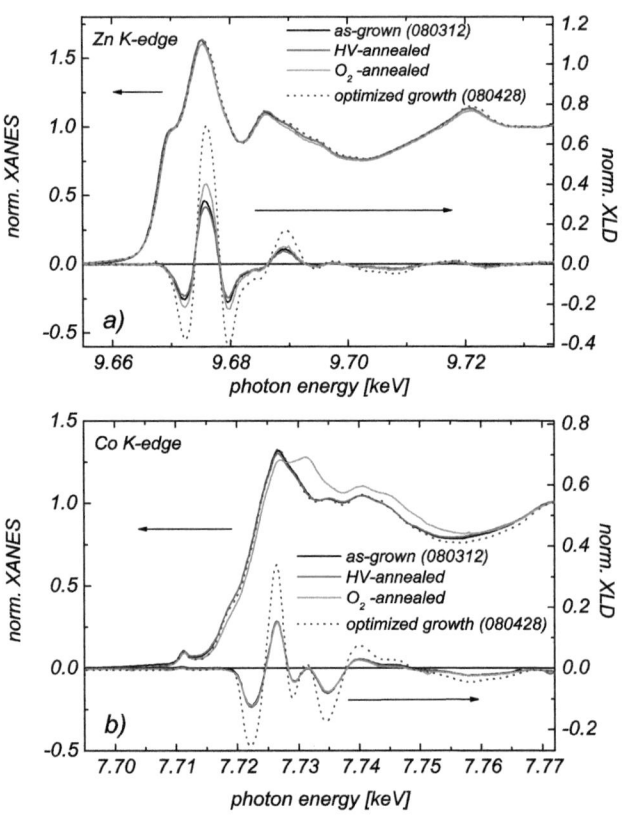

Figure 5.27: Annealing effects observed at the Zn and Co K-edge by XLD of sample 080312. a) A strong structural improvement due to O_2-annealing indicated by an increased signal amplitude of the XLD is visible. The reduction of this amplitude due to HV-annealing is much less pronounced. b) Even though the XANES clearly changes in case of O_2-annealing, the XLD remains almost unchanged. HV-annealing hardly shows any effect for Co in this sample. Note the small absolute value of the XLD. For comparison the data of a sample (080428) of highest structural quality is plotted in addition.

will show the anisotropy of empty electron states characteristic for the ZnO wurtzite lattice. Randomly distributed Co atoms or randomly oriented Co cluster will not cause any signal, since individual anisotropies of electron states will average out and lead to no XLD signature.

Besides random interstitials possible secondary phases have to be considered. If these phases have cubic symmetry, also no XLD signature will be observable. In case of Co mainly three possible combinations should be taken into account: CoO, $ZnCo_2O_4$ and Co_3O_4. The first one crystallizes in cubic structure and was reported to be ferromagnetic in the twenties of the last century [163] but nowadays it is proven to be antiferromagnetic [164]. The second builds a spinel structure including Zn [165]. For this compound distinct magnetization behaviors of bulk and nanoparticles have been recently reported [166]. Similar to CoO the $ZnCo_2O_4$ spinel is also reported to be antiferromagnetic under most conditions [165, 167, 168]. Finally, spinel structured Co_3O_4 is known to form from CoO under elevated temperatures [169]. It is also antiferromagnetic with $T_N = 40$ K [169].

Figure 5.28 compiles XANES spectra of a sample (080428) of best structural quality and of a sample 080312a(O_2) grown under O_2-deficiency as well as spectra of different compounds, namely CoO and Co_3O_4. The absorption spectra of these phases were kindly provided by the group of Prof. Winterer [11] and measured in Argonne Nation laboratory on beamline 12.BM.B. A XANES spectrum of the $ZnCo_2O_4$ compound was not available. However, the formation of the $ZnCo_2O_4$ spinel after O_2-annealing is rather unlikely, since the XLD at the Zn K-edge in figure 5.27 a) is clearly increased, proving more Zn atoms placed on the cation site in proper ZnO wurtzite structure.

From the absorption spectrum it can be concluded that the O_2-annealed sample contains a fraction of Co atoms in an altered phase than before the annealing treatment. The absorption maximum is shifted about 3 eV to a higher energy value. This is known from valence changes of Co and is reported for example in [170]. Obviously the second, annealing related, peak position agrees well with the maximum of the absorption spectrum of Co_3O_4. Co_3O_4 nanoparticles are known to form from CoO already at temperatures of about 300 K [171, 172]. To visualize the contributions of both phases a weighted superposition of the Co_3O_4 spectrum and the spectrum

---

[11]Markus Winterer, Nanoparticle Process Technology, Institute for Combustion and Gas Dynamics, University of Duisburg-Essen, Lotharstrasse 1, D-47057 Duisburg, Germany

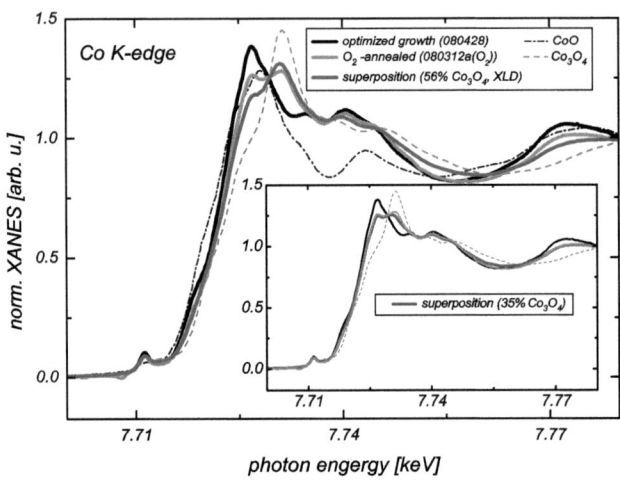

Figure 5.28: Indication for Co_3O_4 spinel by XANES. The red curve is a superposition of the spectrum of sample 080428 and the Co_3O_4 spectrum weighted by the XLD amplitudes. The inset shows a superposition with an optimized weight of the Co_3O_4 spectrum.

of Co:ZnO (080428) of best structural quality is plotted in addition (red curve). The weight of the spectrum of sample 080428 is derived from the ratio of the XLD amplitudes (0.44), assuming the cubic spinel phase of Co_3O_4 does not cause any additional XLD. Even though the height of the two maxima visible in the annealed sample is not exactly matched by the superposition the overall XANES spectrum is reproduced well. Considering the residual deviations an increase of the weight of the spectrum of sample 080428 is tempting to obtain a better fit. The inset of figure 5.28 shows a superposition with a Co_3O_4 spectrum contribution reduced from 56% to 35% yielding an almost perfect fit of the experimental spectrum of the O_2-annealed sample 080312. Note that significant fractions of CoO or Co metal can be excluded, since they inevitably would cause deviations at either the preedge feature or around 7.74 keV, respectively.

ANNEALING EFFECTS

With regard to an optimized weight of 65% of the spectrum of the sample 080428, the reduction of the XLD amplitude of sample 080312a(O_2) at the Co K-edge to 44% of the maximum value for high structural quality has to be scrutinized. Apart from Co atoms bound in secondary phases with no XLD signal, rotated grains can reduce the XLD, as well. Two by 90° [12] rotated crystal grains with similar substitutional Co content will show no XLD, since the XLD of well incorporated Co^{2+} ions will average out.

Even though the samples under investigation are grown epitaxially, oxygen deficient growth reduces the structural quality (figure 5.11). Therefore, a reduction of the XLD due to tilted grains will be present. This means that the XLD amplitude just gives a lower limit of the fraction of the respective phase.

This becomes clearer in the case of the HV-annealed sample (080312a(HV)), which is shown in figure 5.29. If the ferromagnetic-like response known from SQUID measurements (figure 5.25) is attributed to metallic Co cluster, the fraction of Co in this phase can be tentatively weighted by the missing Co K-edge XLD amplitude. Randomly oriented Co metal cluster are not expected to yield any XLD but to affect the XANES – in particular the shape of the preedge feature. From the ratio of the XLD at the Co K-edge of the HV-annealed sample (080312a(HV)) to the ideal XLD one derives 43% of perfect incorporated Co^{2+}. This in turn would mean 57% of metallic cobalt. The resulting XANES spectrum of the superposition of the two phases is plotted in figure 5.29 (green curve). The measured spectrum is hardly matched by the weighted superposition.

Additionally in figure 5.29 superpositions of XANES spectra with arbitrary weighted Co metal content are potted. The experimental XANES of sample 080312a(HV) is well reproduced for a Co content in the range of 5% to 7.5%. Adding additional contributions from CoO or Co_3O_4 yields no significant improvement.

This result is an order of magnitude lower than the fraction derived from the SQUID analysis (see equation 5.7).

The calculation of the magnetization based on XANES results can be preformed the other way round: If the sample only contains 7% of metallic Co, as indicated by XANES, up to 93% of the Co can be paramagnetic Co^{2+}. Thus the magnetic moment measured by SQUID at T = 300 K and an external magnetic field of 4 T would originate from only 7% of the Co atoms with an atomic magnetic moment of

---

[12] c-axis with respect to the E-vector of the incident synchrotron light

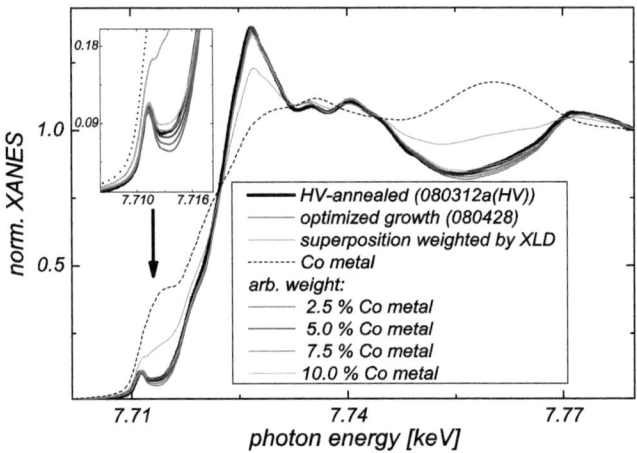

Figure 5.29: XANES analysis of HV-annealed Co:ZnO. A superposition based on the XLD amplitude (green curve) fails to describe the experimental spectrum (bold black). The metallic fraction can be estimated best at the preedge feature (inset).

1.7 μ_B. At T = 5 K and 4 T the paramagnetic Co^{2+}-ions should contribute 80% of their saturation magnetization. In total a magnetic moment of almost $5 \cdot 10^{-7} Am^2$ instead of $5.92 \cdot 10^{-8} Am^2$ should be measured by the SQUID.

In the following the seemingly contradicting results of XLD, XANES, and SQUID- will be discussed:

First, the reduced XLD is likely to be mainly caused by rotated grains, since also the Zn K-edge XLD is strongly reduced to below 41% (figure 5.27). In other words: The formation of wurtzite ZnO requires a higher percentage of properly incorporated Zn atoms. Consistently the XANES spectrum of the sample 080312a(HV) hardly differs from samples grown under optimized conditions.

Second, the discrepancy between the percentage of metallic cobalt in this sample derived by XANES and SQUID analysis has to be explained. The structural investigation method, XANES, is sensitive to all Co atoms, whereas the magnetic method, SQUID, is only sensitive to magnetically active Co atoms. Compensated

ANNEALING EFFECTS

spins of antiferromagnetically coupled Co^{2+}-ions will not contribute to the SQUID signal. Thus the large difference of the fraction of metallic Co, when calculated from SQUID and XANES proves a high amount of compensated magnetic moments.

This is consistent with results for samples of high structural perfection discussed in section 5.2. There the magnetic moment of a sample containing at least 95% of substitutional Co^{2+} is shown to be reduced to $\approx 1/3$ [24]. For the calculation in section 5.2 the absolute amount of Co atoms is used.

In case of sample 080312a(HV) the superparamagnetic fraction of Co atoms, which show a metallic XANES serve as reference for the estimation of the paramagnetic fraction. Thus the result of section 5.2 can be qualitatively confirmed.

Assuming two next cation neighbor Co^{2+}-ions to compensate their moments, i.e. to couple antiferromagnetic via superexchange, none or much reduced/frustrated moments can also be expected for triple and higher order configurations. Taking into account the statistical probability in a 10% Co doped sample (see section 5.2) to find substitutional Co atoms without next Co cation neighbor of 28%, the metallic Co fraction reduces to:

$$\frac{5.21\,Am^2}{0.8 \cdot 0.28} \approx 23.3 \rightarrow \frac{1.75\,Am^2/1.7}{23.3\,Am^2/4.1} \approx 18\% \text{ metallic Co} \qquad (5.8)$$

This result is still more than two times than what is expected from the XANES analysis.

However, some imponderabilities of the SQUID and XANES analysis have to be discussed: First of all the rigid numbers of the atomic moment per Co atom for the paramagnetic ion and the metal have to be questioned. Thinking of small clusters of metallic Co, there will be a large number of interface atoms with reduced quenching of the orbital moment [173]. A second point to mention is the dopant concentration; According to Behringer's calculation (see figure 1.5) the probability of a substitutional Co without next Co neighbor is: $(1-p)^{12}$. Obviously variations of the concentration p result in a different fraction of Co atoms with compensated moments. Note that, in contrast to the calculation in section 5.2 the accuracy of sample size and layer thickness does not affect the calculation as long as no self-absorption perturbs the XANES analysis. For the XANES analysis the normalization of the spectra has to be considered. With regard to the inset of figure 5.29 the error bar can be estimated to be at most 5%.

Despite these uncertainties a large mismatch of a metallic fraction derived from

SQUID magnetometry and XANES analysis is beyond doubt: Note that equation 5.8 - ignoring different atomic moments of metallic and paramagnetic Co atoms - already leads to a four times larger metallic Co fraction of the SQUID analysis.
Finally some important conclusions can be drawn from these considerations: The large mismatch of magnetization contribution of metallic cobalt from SQUID magnetometry and XANES analysis requires a large fraction of compensated Co moments. The compensation cannot be due to fractions of cubic CoO. Due to the six times coordinated lattice site of Co in cubic CoO the valence of Co is different from the four time coordinated wurtzite site. Consistently the Co K-edge XANES of cubic CoO differs significantly from Co^{2+} embedded in wurtzite ZnO. Thus a high fraction of cubic CoO in sample 080312a(HV) can be excluded, since it would change the XANES spectrum. Such a change was discussed previously for the O_2-annealed sample, which revealed a Co_3O_4 fraction. A similar change of the XANES spectrum can be presumed for a possible fraction of $ZnCo_2O_4$, due to the higher valence of the Co atoms [170]. Therefore the compensated Co moments have to be on Zn-cation lattice sites in a ZnO wurtzite environment. Recently published results of wurtzite type CoO nanoclusters embedded in ZnO can explain why such a phase separation is not detected by XRD [174]. Wurtzite type CoO is reported to be magnetically inactive [175, 176], thus suitable to explain discrepancies concerning the amount of metallic Co in sample 080312a(HV) derived from SQUID and XANES analysis. Note that the previous discussion of compensated moments (see also section 5.2) due to a statistical distribution of the Co^{2+}-ions on cation sites fades to the discussion of phase separated CoO in wurtzite structure.

In the final part of this section the results of magnetic resonance measurements of the HV-annealed samples shall be addressed. The presentation of the results will be mainly limited to a phenomenological description. A possible explanation of the signal will be given in section 5.6.

In figure 5.30 the change of the ESR spectrum of a 10% Co doped sample (080122) at T = 300 K after a HV-annealing procedure (450 °C, 30 minutes) is shown. Note the large field range of the spectra from 0 mT to 600 mT. Figure 5.30 a) suggests an extremely broadened line in the spectrum of the HV-annealed sample. Superimposed one can find: i) the paramagnetic centers of the substrate which illustrate that the samples were identically oriented for both scans, ii) the broad central resonance around g = 2 ($\approx$ 337 mT) which has been discussed in a prior section (5.4) and

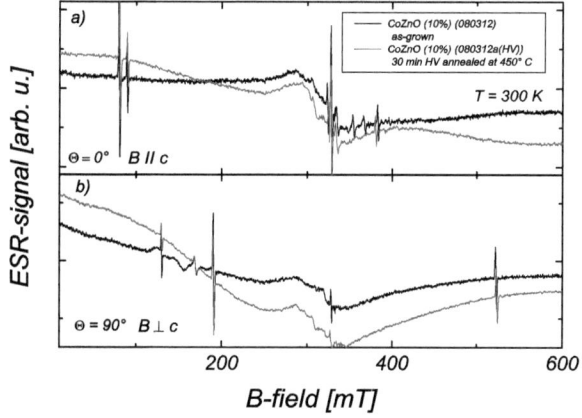

Figure 5.30: Altered magnetic resonance spectrum of sample 080312 after HV-annealing. No additional resonance signal is obvious but a changed shape of the total spectrum. Narrow resonance signals belong to paramagnetic signals of the substrate and prove the proper alignment of the samples.

is assigned to superparamagnetic cluster ensembles. The lower plot b) is taken in the perpendicular geometry. At fields up to ≈ 300 mT the annealed sample shows a more pronounced negative slope, which is very faint also visible in the as-grown sample. At higher fields the gradient of the spectrum of the annealed sample is positive in b) in contrast to the $\theta = 0$ spectrum. These phenomenological results are representative for the angular dependence of the ESR spectra of HV-annealed samples grown at reduced O_2 pressure: A change of the shape of the spectra can be observed but an ordinary angular dependency of an additional magnetic resonance line is not visible. In particular no crossing of the g = 2 ("$\frac{\omega}{\gamma}$") field value (≈ 337 mT) can be observed. The HV-annealing procedure seems to enhance a behavior which is already present in the as-grown sample, since slight changes of the slope are also observable in the as-grown sample but much less pronounced.

Figure 5.31 shows an alternative evaluation of the data obtained from sample 080122 and its HV-annealed version at T = 5 K. Generally the zero line during the

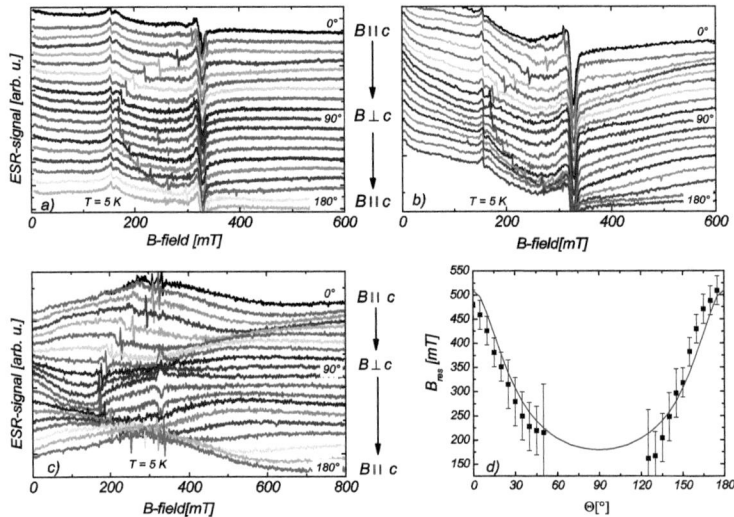

Figure 5.31: a) Angular dependent spectra measured of the as-grown sample 080122. All visible resonances can be found in the sapphire substrate, cavity or cryostat. b) Similar measurement after HV-annealing. A clear change in the shape of the spectra is visible. Superimposed one can guess a faint angular dependent feature. c) Residual signal after subtraction of an averaged signal. A broad resonance with a uniaxial angular dependence becomes visible. d) Angular dependent resonance fields of the signal in c). The line shape varies very much and the line vanishes partly in the zero field. The given error bars are estimated. The blue curve is a guide to the eye.

ANNEALING EFFECTS

magnetic resonance measurements was satisfactory, besides the well known features from sample holder and cavity. Therefore a background subtraction was often obsolete. A respective plot of the as-grown sample is shown in figure 5.31 a). The HV-annealed samples turned out to be different in that respect, since they reveal a strongly altered background which changes only slightly with the sample angle as shown in b) for sample 081122a(HV). The change is similar to what was discussed before (figure 5.30) at a temperature of 300 K. In particular the resonance curve at lower fields shows a steady negative slope from 0 mT up to roughly 350 mT. A steady background drift can be excluded, due to the signal at high field values, which is again close to the zero line. To derive further information and to reveal eventual hidden signals an isotropic background signal was subtracted from the data of plot b). The respective curve was generated by averaging over all scans. After subtraction of this averaged spectrum, angular dependent features in the spectra are expected to be pronounced due to suppressed isotropic contributions. In part c) of figure 5.31 the derived residual angular dependent features are shown. Besides the narrow paramagnetic resonance from the sapphire substrate, a very broad line with clear uniaxial angular dependence is visible. The FWHM can be estimated to be about 300 mT, but since the line shape varies with the angle one has to assume a large uncertainty. In figure d) respective resonance field values are plotted for angles from $\theta = 0°$ to $180°$. The missing data points from $\theta = 60°$ to $120°$ result from the fading of the line and possibly partly disappearance in zero field. The blue curve is a guide to the eye. The changing size of the given error bars reflect the fading of the signal from $60°$ to $120°$.

The similar procedure as described above has been applied to the data of the as-grown sample. The resulting residual spectra did not reveal a credible angular dependent resonance signal.

The strong angular dependence of the line shape itself in figure 5.31 suggests a superposition of several individual components with resonance fields in the respective field range. The same analysis was used for the spectra taken at $T = 300$ K. Also in this case a clear angular dependent signal can be derived (not shown) but the shape of the resonance curve is poor. It is neither a Gaussian nor Lorentzian and changes strongly with the angle - much more than for the $T = 5$ K data.

Better results at $T = 300$ K were obtained for the sample 080312 which was also annealed similarly under high vacuum conditions. Its magnetization behavior has

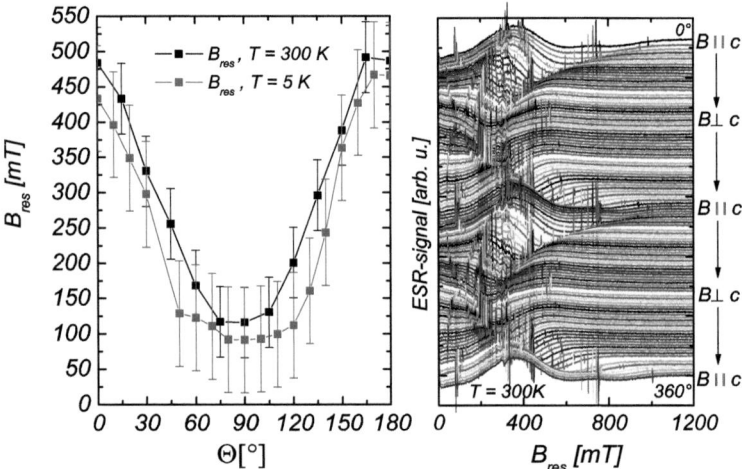

Figure 5.32: Magnetic resonance signal in sample 080312 after HV-annealing and subtraction of an averaged background. The right plot shows the very detailed angular measurements at T = 300 K, clearly revealing a uniaxial dependency of the broad resonance. Narrow paramagnetic resonances of the substrate are superimposed. The left plot shows exemplarily derived values of the resonance field for the measurement shown on the right (black symbols). Low temperature values are additionally plotted (red symbols). Due to the heavily distorted shape of the resonance features error bars up to 50 mT are shown.

already been presented in figure 5.25 and is qualitative similar to sample 080122. The very detailed angular dependent scan of spectra after subtraction of an averaged background is shown in figure 5.32 on the right. The left part of the figure shows exemplarily values of the resonance fields. Due to the poor line shape error bars for the resonance field up to 50 mT are assumed.

The angular dependent spectra taken at T = 5 K of the HV-annealed version of sample 080312 - analyzed similarly - also reveal a uniaxial broad resonance (not shown). A uniaxial shift of a resonance with a FWHM of about 300 mT is visible. The respective resonance field values are additionally plotted (red symbols) in the left part of figure 5.32.

Assuming a ferromagnetic behavior one should expect the anisotropy to increase with decreasing temperature. This would mean higher resonance fields in the out-of-plane case and lower values for the in-plane case. Instead the low temperature resonance fields seem to be systematically at lower field values. However, within the error bars no final conclusion can be made. The as-grown version of sample 080312 was used for a similar analysis but no angular dependent features were found.

In summary, after HV-annealing the samples 080122 and 080312 show a very broad magnetic resonance from T = 5 K up to T = 300 K superimposed on an isotropic spectrum. The high uniaxial anisotropy and the temperature dependence of the resonance field are in agreement with an enhanced fraction of metallic Co in the HV-annealed samples as it was also found by XANES and SQUID analysis.

5.6 (Blocked) Superparamagnetic powder interpretation of isotropic ESR spectra contributions

One of the remaining questions concerning the RMS grown Co:ZnO samples of reduced quality is why magnetic resonance is hardly capable of observing the magnetization which is visible by SQUID-magnetometry. As shown in section 5.5 a uniaxial magnetic resonance in HV-annealed samples becomes visible only after subtraction of an isotropic signal. This section will try to give an explanation for this isotropic signal.

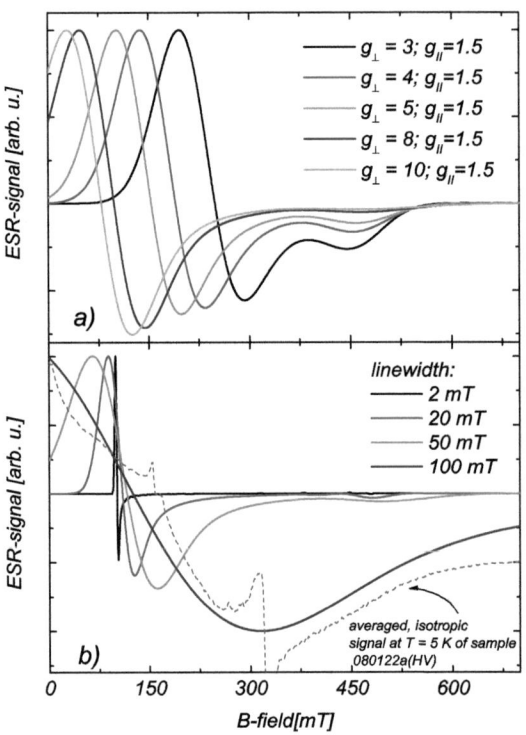

Figure 5.33: Simulation of powder spectra. The upper plot shows the effect of increasing anisotropy of the paramagnetic centers by varying $g_\perp$ and $g_\parallel$. In the lower plot solely the linewidth was changed.

Figure 5.33 shows various simulated powder spectra, as introduced in section 1.3.2. The simulations were calculated with the ESR simulation software package XSOPHE [177], which performs a numerical integration according to equation 1.34. The spectral shape develops due to the random distribution of orientations of the crystallites. The powder spectrum represents a weighted superposition of all possible resonance fields, since the resonance field of paramagnetic centers depends on the orientation of the crystallites.

In the upper plot a) the anisotropy has been varied by different $g_{\perp/\parallel}$. As already shown in figure 1.3 for high anisotropies the low resonance field value becomes dominating and the line shape looks almost like the first derivative of a Gaussian. For smaller anisotropies the negative dip at low fields becomes damped and a dip at the higher resonance field evolves.

In the lower part of figure 5.33 the linewidth of the individual signal was varied. The lower and upper resonance fields were kept constant at values of 150 mT and 500 mT, which correspond to g-values of 6.7 and 1.4 at X-band frequency. For large linewidths and large anisotropies the spectra can partly disappear in zero field. In case of narrow linewidths the resonance fields can be clearly identified by the peaks or dips, respectively.

Additionally, in figure 5.33 b) the averaged low temperature ESR spectrum of a high vacuum annealed 10% doped Co:ZnO sample (080122a(HV)) is plotted. As described in section 5.5 this spectrum was subtracted from individual spectra to identify angular dependent features. In turn, this means that the curve derived from averaging over all angles emphasizes isotropic features. Disregarding the background signal of the cavity (around 330 mT) and a broad peak (around 150 mT) typical for sapphire substrates the spectrum qualitatively resembles a powder spectrum.

This similarity can be understood considering the derivation of the powder spectrum (chapter 1). In equation 1.34 the resonance condition of a paramagnetic center was inserted. In principle this simply reflects a magnetic moment with a hard- and easy axis — a description which can also account for nanoclusters. Even without explicitly solving the Landau-Lifshitz equation, the angular dependence of the resonance condition of a nanoparticle can be understood as follows:

If there is no easy axis the magnetization direction of a nanoparticle can be oriented arbitrarily. In case of an external field the magnetization will be oriented along the

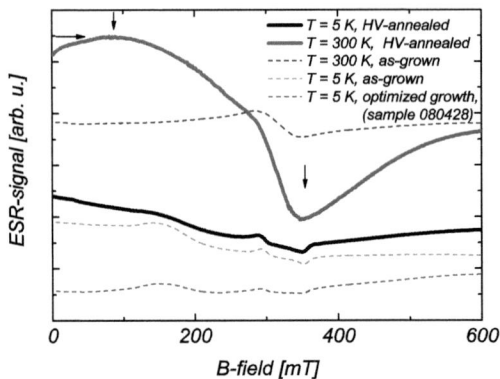

Figure 5.34: Comparison of isotropic ESR signals. After HV annealing the spectra show a much pronounced slope. At T = 300 K a maximum is present in the field range, indicative for reduced anisotropy of blocked clusters. Note: spectra are not to scale.

field direction, even for smallest fields. In principle, the resonance field of all such nanoparticles is similar, i.e. $B_{res} = \frac{\omega}{\gamma}$. This is the case for ideal superparamagnetic nanoparticles.

In case of a magnetic anisotropy, i.e. an easy axis, the situation changes, since an external field has to align the magnetization direction against the intrinsic anisotropy of the nanoparticle. This happens more easily for field directions close to the easy axis than perpendicular. Thus the resonance fields will not change a lot for small angles between easy axis and external field. In turn large angles between easy axis and external field will result in high resonance fields with strong angular dependence.

Thinking of an ensemble of randomly oriented easy axis the latter scenario has all the requirements needed for a powder spectrum, as it is described in chapter 1.

The issue of ferromagnetic resonance of an ensemble of nanoparticles with randomly distributed anisotropy axes has recently been treated in more detail [178].

Figure 5.34 finally corroborates the above considerations. It compiles spectra of HV-annealed and as-grown samples (080122) at temperatures of 5 K and 300 K.

(BLOCKED) SUPERPRAMAGNGETIC POWDER INTERPRETATION

For better visibility the spectra have been smoothed to remove distracting features like paramagnetic impurities. Qualitative comparison of the spectra yields some valuable insights:

Taking into account the above model for a powder spectrum consisting of blocked magnetic moments of clusters one would expect the spectral shape to change with temperature. The anisotropy of ferromagnets is expected to decrease with increasing thermal excitation. In principle this effect is modeled in figure 5.33. The ESR-spectrum taken of the HV-annealed sample (080122) in figure 5.34 at $T = 5$ K (black curve) and $T = 300$ K (red curve) reflects this behavior. The black arrows show in particular: A peak at low fields for $T = 300$ K which is not visible in the $T = 5$ K data (shifted to zero field), an enhanced minimum for the $T = 300$ K curve. While the temperature related shifting of the first peak is understandable straight forward (reduced anisotropy), the enhanced minimum around 350 mT needs additional considerations. It cannot be the minimum occurring due to the high resonance field of the powder spectrum, since one would expect first a dip originating at the low resonance field. The latter in turn is expected to be reduced with smaller anisotropy, i.e. for higher temperatures (see figure 5.33 a)).

The origin of the pronounced minimum becomes evident considering the $T = 300$ K spectrum of the as-grown sample, which shows a central peak originating form superparamagnetic clusters as discussed in section 5.4. Therefore the minimum of the HV annealed sample at $T = 300$ K in figure 5.34 is likely to be the superposition of the low field dip of the powder spectrum and the superparamagnetic resonance. Consequently, the 5 K spectra of annealed and as-grown samples resemble each other much more than the 300 K spectra do, since the superparamagnetic contribution is not. Nevertheless the HV-annealed version of sample 080122 has an enhanced slope and indicates a faint broad minimum around 350 mT. In addition, in figure 5.34 a spectrum of a sample of high structural quality is shown, without any indications of a spectrum of blocked clusters.

These results suggest – in agreement with SQUID and XANES analysis – that the metallic clustering in samples grown at reduced oxygen partial pressure is enhanced by HV-annealing. It might lead to either improved structural properties of the clusters or simply to growth of the clusters, which both lead to improved ferromagnetic-like properties, as e.g. higher T_Cs, of the individual particles.

5.7 Discussion: Co:ZnO

The dilute magnetic semiconductor Co:ZnO was the focus of the present work. A series of samples was grown by RMS. In addition to a structural examination by XRD the material was characterized by synchrotron measurements at the ESRF. The XLD gives valuable insight of the local structural environment of Zn and Co cations. A characteristic XLD pattern of Co^{2+} could be evaluated quantitatively yielding more than 95% of Co atoms substituting Zn in high quality films. The structural quality of the samples grown by RMS is comparable with the best Co:ZnO samples grown by PLD.

A detailed magnetic characterization of high quality ZnO doped with 10% Co by SQUID, ESR, and XMCD was conducted. Contrary to many reports in the literature e.g. [22, 26, 179] no room temperature ferromagnetism was found. Instead paramagnetic behavior of dopant and host has been proven [24]. Magnetic resonance measurements reveal an intense line broadening of the Co^{2+} single ion spectra, presumably caused by dipolar interactions present in Co:ZnO for high concentrations of Co^{2+}.

For a 10% Co doped Co:ZnO sample the effective magnetic moment per Co atom has been determined independently from SQUID and XMCD measurements. The reduced effective magnetic moment derived from SQUID measurements reveals about 70% of the Co moments to be compensated. Assuming a random, statistical distribution of the Co atoms on cation lattice sites, only 28% of the Co atoms are without at least one next Co cation neighbor. An antiferromagnetic coupling between next cation neighbor Co atoms can account for the reduced magnetic moment, since also higher orders of Co next cation neighbor configurations presumably have either compensated or strongly reduced magnetic moments. Thus the magnetic moment of high quality Co:ZnO can be quantitativly understood by considering the statistical concentration of Co atoms on cation lattice sites without next Co cation neighbor. A deviation from the optimal growth condition, e.g., reduced oxygen partial pressure, is shown to not only reduce the structural quality but also to induce ferromagnetic-like signatures. Coercivity and remanence and separation of FC/ZFC curves increase with oxygen deficiency. The reduction of the percentage of substitutional Co, as measured by XLD can be correlated to the onset of ferromagnetic-like signatures. A low temperature maximum of ZFC curves in M(T) measurements indicate blocking of magnetic supermoments (i.e. metallic clusters) as origin of the observed behavior.

Magnetic resonance measurements of Co:ZnO samples grown under oxygen deficiency show similar spectral features as Co/CoO core shell nanoparticles: At low temperature the absence of any angular dependent signal is indicative of magnetically blocked clusters, whereas at room temperature a broad resonance with uniaxial anisotropy is typical for a superparamagnetic ensemble with dipolar coupling.

The shape of an isotropic signal of the low temperature ESR spectra of these samples are shown to be consistent with the perception of a blocked ensemble of superparamagnetic clusters with randomly oriented uniaxial anisotropies.

In the case of one sample grown on r-plane sapphire severe metallic precipitations cause ferromagnetic resonance spectra with high anisotropies down to 5 K. Neither SQUID nor ESR measurements indicate a blocking behavior. Temperature dependent shifts of the resonance field are observed and corroborate a ferromagnetic origin. XANES and XMCD spectra consistently show strong signatures from the metallic Co phase in this sample.

Finally, O_2 and HV-annealing experiments were performed at temperatures up to 450 °C. Structurally high quality Co:ZnO, with Co^{2+} being almost completely on substitutional sites, was shown to be practically inert to HV-annealing.

In samples grown under oxygen deficiency ferromagnetic-like behavior could be fully removed by O_2-annealing. XANES analysis revealed a large fraction of Co_3O_4 after O_2-annealing. HV-annealing in turn resulted in improved ferromagnetic-like properties. Element specific structural investigations show that this is only caused by Co atoms on non substitutional sites. The HV-annealed samples reveal broad magnetic resonance signatures with uniaxial out-of-plane anisotropy superimposed on signals known from magnetically blocked or superparamagnetic clusters. Estimates for a metallic cobalt phase from XANES analysis and SQUID magnetometry yield different results which can be understood by assuming a high fraction of magnetically inert Co, as it was also found in best structured samples. In turn, the altered ferromagnetic-like properties have to be attributed to improved magnetic properties of a metallic Co phase and the respective clusters.

Summary

The two wide band gap DMS Gd:GaN and Co:ZnO were the focus of the present work. Room temperature ferromagnetism, as frequently claimed theoretically and experimentally, is up to now under debate. Clarification is of high technological interest, since this property would make both materials candidates for potential spintronic applications.

Two types of Gd:GaN films were under investigation, fabricated either by FIB implantation of GaN or grown by MBE. Ferromagnetic-like behavior up to room temperature, as it was found by SQUID in implanted films grown on SiC, could not be reproduced with GaN films grown on sapphire implanted with similar Gd doses. Both sets of samples are of comparable structural quality. Within this work, no conclusive explanation for these contradictory results could be found.

Synchrotron measurements on a MBE grown sample, which indicate room temperature ferromagnetism as measured by SQUID, yield element specific insight into the magnetic properties. An s-shape, (super-)paramagnetic-like M(H) curve measured by XMCD at the Gd-L_3 edge reveals the magnetic behavior of the Gd dopant to be independent of the ferromagnetic-like integral magnetic behavior.

In addition, a possible host matrix polarization has been investigated. XMCD measurements at the Ga K-edge reveal a very small polarization, which is at least one order of magnitude too low to account for the total magnetization. Low temperature ESR spectra indicate a fraction of ferromagnetic nanoclusters, probably composed of phase separated Gd and GdN. The influence of these phases on the element specific M(H) curves can be described by superimposed Brillouin and Langevin functions.

In summary ESR, XMCD and a majority of SQUID measurements cannot corroborate room temperature ferromagnetism in Gd:GaN. Instead paramagnetic behavior of the dopant is found. In view of contradictory SQUID measurements on some samples, which suggest ferromagnetism at room temperature, two points should be considered: i) Sample contamination during growth or afterwards. ii) Artifacts caused by the SQUID magnetometer.

Co:ZnO was predominantly grown by reactive magnetron sputtering in a newly build UHV system. Composite sputter targets with 90% Zn and 10% Co resulted in Co:ZnO samples with a dopant concentration close to 10%. XLD measurements show 95% of the Co to be incorporated on Zn substitutional sites as Co^{2+}, in case of optimized growth conditions. These samples show a paramagnetic like behavior down to 5 K which can be described by a Brillouin function.

A reduced total magnetization as measured by SQUID can be explained by compensated moments due to antiferromagnetic coupling of Co moments occuring according to random statistical dopant distribution on cation sites. ESR spectra of Co^{2+} in these samples are intensely broadened due to dipolar interactions caused by the high dopant concentration.

RMS growth of Co:ZnO under reduced oxygen partial pressure is shown to lead to magnetic signatures typical for phase separated clusters. SQUID measurements reveal FC/ZFC separation and M(H) hysteresis loops up to room temperature. Maxima in ZFC measurements indicate blocking of magnetic (super-)moments. ESR signals are observed only at elevated temperatures (> 60 K) which is typical for ensembles of unblocked superparamagnetic nanoparticles. Depending on the density of superparamagnetic moments a uniaxial anisotropy occurs due to dipolar interaction. XLD consistently shows a reduced fraction of substitutional Co. XANES and XMCD measurements finally corroborate a enhanced metallicity in these samples. *Therefore the ferromagnetic-like properties are due to extrinsic origins and not intrinsic to this material.*

Further clarification of the nature of phase separation in Co:ZnO was gained by annealing experiments. The samples grown in oxygen deficient atmosphere were heated up to 450° C either under HV or O_2 atmosphere. In case of HV-annealing a strong increase of the ferromagnetic-like properties, as measured by ESR and SQUID, is accompanied by hardly any change of the XLD at the Co K-edge. Therefore the ferromagnetic-like properties of these Co:ZnO films are governed by non-substitutional Co atoms. O_2-annealing in turn resulted in a collapse of the ferromagnetic properties. As can be deduced from XANES spectra the formation of the Co_3O_4 spinel is induced. Also after this treatment the XLD at the Co K-edge is hardly affected. Samples of high structural quality were also subject to annealing

Summary

procedures. Hardly any effect on the magnetic and structural properties was observed. These results prove the substitutional Co^{2+} ions to be practically inert to the annealing procedure, whereas non-substitutional Co is strongly affected.

Considering the results of this work ferromagnetic-like behavior at least in 10% doped Co:ZnO is caused by secondary phases, whereas the intrinsic magnetic property of the homogeneous system is paramagnetism.

Outlook

Ferromagnetic signatures are found in both Gd:GaN and Co:ZnO. Even though the origin of the spin polarization is not intrinsic, the technological relevant carrier spin polarization might be achieved. The findings of this work offer recipes to tailor properties of precipitations far below the detection limit of conventional XRD analysis. Scattering events in such two phase materials will counteract the persistence of spin polarization, therefore a spin polarization will probably depend on cluster sizes and distributions. Another important aspect concerning all kinds of DMS is co-doping. In particular ZnO is known to be intrinsically n-type, which is valid for all samples of this work. The role of hydrogen as shallow donor has recently come into focus of research efforts [180]. Theoretical claims of hole-mediated long range magnetic order are difficult to verify experimentally, since p-type ZnO has been rarely produced so far. Most reports on p-type ZnO are based on N-doping, but also As or P seem to be feasible as dopant [43]. p-type Co:ZnO is certainly worth to be investigated in the future. Moreover the use of co-doping to create metallic-like n-type samples might be promising, since ferromagnetism has been reported to evolve from carrier concentrations above 10^{20} electrons/cm^3 [181]. Carrier concentrations of Co:ZnO sample investigated in this work are in the rage from 10^{17} to 10^{19} and no ferromagnetic transport signatures are found [182].

The goal of ESR measurements within this work was predominantly to identify additional magnetic phases after incorporation of the dopant in ZnO or GaN, respectively. This was successfully achieved in particular for the Co^{2+}-ion, blocked clusters with supermoments, and unblocked supermoments, partly showing dipolar interactions. ESR can also offer valuable information concerning the interactions by analyzing the line shape. Future studies should to be conducted on either much thicker films, containing more material, or samples with lower dopant concentration, since intense line broadening aggravates measurements on 10% doped Co:ZnO.

Such samples would also be suitable for multi-frequency ESR measurements for e.g. thorough g-factor studies.

Moreover the role of phase separated CoO and Co_3O_4 should be investigated. In particular CoO is of interest, since it has been recently shown by synchrotron measurements to exist also in the wurtzite phase [174]. Up to now the knowledge about nanoscopic clusters of these phases embedded in a crystal matrix is very limited and the respective investigations remain challenging.

Appendix A

Calculation of the Co^{2+}-ion in ZnO

Starting from the free d^7-ion the calculation of Co^{2+} in ZnO first needs to consider the different orders of magnitude of the energies of the Hamiltonian. The

$$H_{Coulomb} \approx 10^0 eV > H_{CF-tetra.} \approx 10^{-1} eV > H_{CF-trig.} \approx 10^{-2} eV$$
$$> H_{S-O} \approx 10^{-3} eV > H_{Zeeman} \approx 10^{-4} eV$$

second and third term are treated separately by group theory. Figure A.1 shows the description of the tetrahedrally bound cobalt within a cube, assuming a perfect tetrahedron. If the coordinate system is chosen that way that the Co^{2+} is placed in the origin, the ligand positions are given by (a,-a,a), (a,a,-a),(-a,a,a) and (-a,-a,-a), with $\sqrt{3a^2} = d$=next neighbor distance. Within the point charge approximation the resulting potential can be calculated by summing up the Coulomb contributions of the four ligands. A conversion of this potential into spherical harmonics enables the description of the wave functions of the free ion and the potential by the same basis.

The symmetry of the potential is referred to as T$_d$. It consists of 8 rotations around a threefold axis (C$_3$), two rotations around a twofold axis (C$_2$), 6 rotation reflections (S$_4$) and 6 diagonal reflexions (σ_d). This symmetry is then compared to the symmetry of the ^{4}F degenerate ground state of the free ion. It can be shown by group theory [183] that this symmetry properties of the potential lead to a splitting of the degenerate energy levels of the free ion. In particular the irreducible representation yields one orbital singlet (A$_2$) and two triplet states (T$_1$ and T$_2$).

The Co^{2+} is not perfectly tetrahedrally bound, but slightly contracted along the c-axis. Therefore a trigonal ligand field component has to be considered. The symmetry of the trigonal distortion is described by a C$_{3v}$ point group. It solely affects

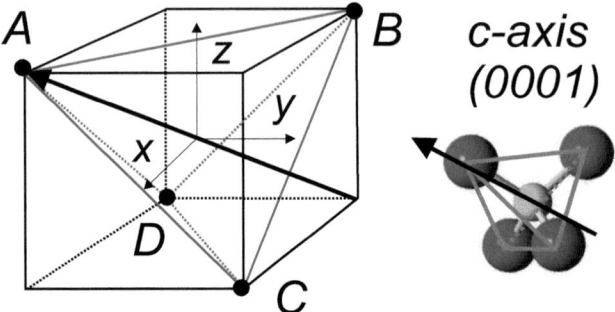

Figure A.1: Tetrahedral crystal field description

the triplet states and leads to splitting into doublet and singlet states [184].

Finally the last application of the group theory concerns the spin-orbit coupling. For this the double group concept by addition of the spinor has to be used. As depicted in figure 1.2 this reduces the degeneracy of the Co^{2+}-ion in ZnO to 14 solely spin degenerate states

For magnetic resonance measurements only the lowest two states – both Kramers doublets – are relevant.

The standard treatment of 3d-ions in an intermediate crystal field would be to apply perturbation theory according to the orders of magnitude as it is given in A. However in case of Co^{2+} in ZnO the spin-orbit coupling and the Zeeman splitting are of the same order of magnitude. In particular the splitting of the S=1/2 and S=3/2 lowest energy Kramers doublets are only 0.7 meV separated [55]. Respective problems and modifications for the application of perturbation theory are mentioned by [185, 186].

A proper derivation of the spin Hamiltonian $H_{Spin} = H_{S-O} + H_{Zeeman}$ by perturbation theory yields[187]:

$$H_{spin} = \mu_B g_\parallel H_z S_z + \mu_B g_\perp (H_x S_x + H_y S_y) + DS_z^2 - \frac{1}{3}DS(S+1) + \ldots \quad (A.1)$$

The last term $\frac{1}{3}DS(S+1)+\ldots$ and all further terms simply causes constant energy shifts. The most common form of the Hamiltonian to describe Co^{2+} in ZnO is therefore given by:

$$H_{spin} = \mu_B g_\parallel H_z S_z + \mu_B g_\perp (H_x S_x + H_y S_y) + DS_z^2 \quad (A.2)$$

APPENDIX D: CALCULATION OF THE Co^{2+}-ION IN ZnO

The two g-factors $g_\parallel$ and $g_\perp$ and the zero field splitting D therefore suffice to describe the system. These values contain the influence of the crystal field and the spin-orbit coupling. A detailed calculation yields (see [54]):

$$D = \frac{\lambda^2}{\Delta^2}[2\nu - \frac{10\sqrt{2}}{3}\nu'(1+\frac{4\Delta}{75B})]$$
$$g_\parallel = g_e - \frac{8\lambda k}{\Delta}[1 - \frac{\nu}{3\Delta} + \frac{5\sqrt{2}}{9\nu'}(1+\frac{4}{75}\frac{\Delta}{B})] \quad (A.3)$$
$$g_\perp = g_e - \frac{8\lambda k}{\Delta}[1 + \frac{\nu}{6\Delta} - \frac{5\sqrt{2}}{18\nu'}(1+\frac{4}{75}\frac{\Delta}{B})]$$

In these formulas a reduction of the orbital coupling parameter λ is accounted for by the dimensionless parameter k, the isotropic g value is denoted by g_e and Δ, ν and ν' are the parameter for the description of the crystal field [188]. B is the Racah parameter [189]. If the external magnetic field is applied along the z-direction A.2 is a diagonal matrix.

$$\begin{bmatrix} \frac{9}{4}D - \frac{3}{2}\mu_B g_\parallel H_z & 0 & 0 & 0 \\ 0 & \frac{1}{4}D - \frac{1}{2}\mu_B g_\parallel H_z & 0 & 0 \\ 0 & 0 & \frac{1}{4}D + \frac{1}{2}\mu_B g_\parallel H_z & 0 \\ 0 & 0 & 0 & \frac{9}{4}D + \frac{3}{2}\mu_B g_\parallel H_z \end{bmatrix}$$

The respective eigenstates are described by the quantum number m_s=-3/2...+3/2. The energy values are derived straightforward by:

$$E(-3/2) = \frac{9D}{4} - \frac{3}{2}\mu_B g_\parallel H_z$$
$$E(-1/2) = \frac{D}{4} - \frac{1}{2}\mu_B g_\parallel H_z$$
$$E(+1/2) = \frac{D}{4} + \frac{1}{2}\mu_B g_\parallel H_z \quad (A.4)$$
$$E(+3/2) = \frac{9D}{4} + \frac{3}{2}\mu_B g_\parallel H_z$$

As one can see from figure A.3 for high magnetic field the ground state of the ion will change form the m_s=-1/2 to the m_s=-3/2. At sufficient low temperatures and/or sufficient high external fields a respective magnetization step is observable [190].

If the c axis is oriented perpendicular to the external field the Hamiltonian A.2 contains non diagonal elements due to the application of ladder operators [1].
Also this matrix can be diagonalized yielding the respective energy levels:

---

[1] $S_x = \frac{S_+ + S_-}{2}, S_y = \frac{S_+ - S_-}{2i}$

APPENDIX D: CALCULATION OF THE Co^{2+}-ION IN ZnO

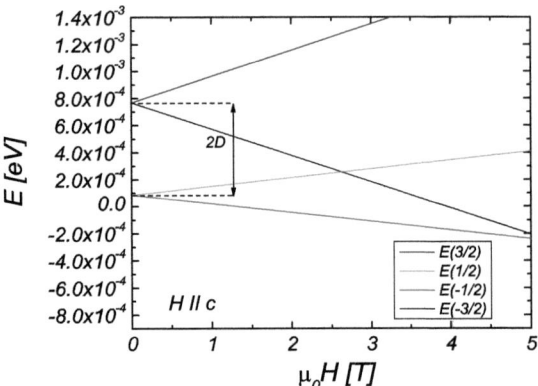

Figure A.2: Energy levels of Co^{2+} for $H \parallel c$. Energy values refer to the case D=0.

$$\begin{bmatrix} \frac{9}{4}D & \frac{1}{2}\sqrt{3}\mu_B\, g_\perp\, H_x & 0 & 0 \\ \frac{1}{2}\sqrt{3}\mu_B\, g_\perp\, H_x & \frac{1}{4}D & \mu_B\, g_\perp\, H_x & 0 \\ 0 & \mu_B\, g_\perp\, H_x & \frac{1}{4}D & \frac{1}{2}\sqrt{3}\mu_B\, g_\perp\, H_x \\ 0 & 0 & \frac{1}{2}\sqrt{3}\mu_B\, g_\perp\, H_x & \frac{9}{4}D \end{bmatrix}$$

$$\begin{aligned} E(4) &= \tfrac{1}{2}\mu_B\, g_\perp\, H_x + \tfrac{5}{4}D + \sqrt{\mu_B^2 g_\perp^2 H_x^2 - D\, g_\perp\, H_x\, \mu_B + D^2} \\ E(3) &= -\tfrac{1}{2}\mu_B\, g_\perp\, H_x + \tfrac{5}{4}D + \sqrt{\mu_B^2 g_\perp^2 H_x^2 + D\, g_\perp\, H_x\, \mu_B + D^2} \\ E(2) &= \tfrac{1}{2}\mu_B\, g_\perp\, H_x + \tfrac{5}{4}D - \sqrt{\mu_B^2 g_\perp^2 H_x^2 - D\, g_\perp\, H_x\, \mu_B + D^2} \\ E(1) &= -\tfrac{1}{2}\mu_B\, g_\perp\, H_x + \tfrac{5}{4}D - \sqrt{\mu_B^2 g_\perp^2 H_x^2 + D\, g_\perp\, H_x\, \mu_B + D^2} \end{aligned} \quad (A.5)$$

APPENDIX D: CALCULATION OF THE Co^{2+}-ION IN ZnO

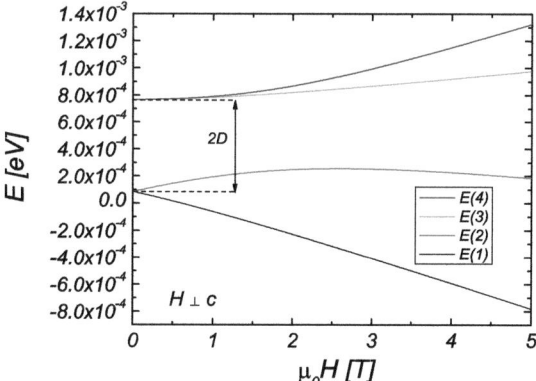

Figure A.3: Energy levels of Co^{2+} for $H \perp c$. Energy values refer to the case D=0.

Appendix B

Sample overview

type/ signature	Gd content	substrate	structure orientation	growth	annealing
Gd:GaN J0174	2.9%	Al_2O_3 (sapphire)	wurtzite (0001)	MBE, N-plasma assisted	
Gd:GaN 2492	nom. $2 \cdot 10^{19}$ Gd/cm^3	6H-SiC	wurtzite (0001)	MBE, NH_3-ass. assisted	
Gd:GaN 2570-3	$1 \cdot 10^{20}$ Gd/cm^3	6H-SiC	wurtzite (0001)	MBE, NH_3-ass., Gd^{3+} FIB	
Gd:GaN 2570-3a	$1 \cdot 10^{20}$ Gd/cm^3	6H-SiC	wurtzite (0001)	MBE, NH_3-ass., Gd^{3+} FIB	rap. therm. 900°C, 30 sec.
Gd:GaN 2570-2	$1 \cdot 10^{18}$ Gd/cm^3	6H-SiC	wurtzite (0001)	MBE, NH_3-ass., Gd^{3+}FIB	
Gd:GaN 2570-1	$1 \cdot 10^{16}$ Gd/cm^3	6H-SiC	wurtzite (0001)	MBE, NH_3-ass., Gd^{3+}FIB	
GaN 2570-0	-	6H-SiC	wurtzite (0001)	MBE, NH_3-ass.,	
Gd:GaN N736h	$1 \cdot 10^{20}$ Gd/cm^3	Al_2O_3 (sapphire)	wurtzite (0001)	MBE, NH_3-ass., Gd^{3+}FIB	
Gd:GaN N736i	$1 \cdot 10^{19}$ Gd/cm^3	Al_2O_3 (sapphire)	wurtzite (0001)	MBE, NH_3-ass., Gd^{3+}FIB	
GaN N736	-	Al_2O_3 (sapphire)	wurtzite (0001)	MBE, NH_3-ass., Si:GaN buffer	

type/ signature	Co content	substrate	structure	growth (Ar:O$_2$)	annealing
Co:ZnO 080122	nomin. 10%	Al$_2$O$_3$ (sapphire)	wurtzite (0001)	RMS (reac. mag. sputtering)(10:0.5)	
Co:ZnO 080122a(HV)	nomin. 10%	Al$_2$O$_3$ (sapphire)	wurtzite (0001)	RMS (10:0.5)	HV, 450°C 30 min.
Co:ZnO 080122a(O$_2$)	nomin. 10%	Al$_2$O$_3$ (sapphire)	wurtzite (0001)	RMS (10:0.5)	O$_2$, 450°C 30 min.
Co:ZnO 080124	nomin. 10%	Al$_2$O$_3$ (sapphire)	wurtzite (0001)	RMS (10:0.4)	
Co:ZnO 080128	nomin. 10%	Al$_2$O$_3$ (sapphire)	wurtzite (0001)	RMS (10:1)	
Co:ZnO 080312	nomin. 10%	Al$_2$O$_3$ (sapphire)	wurtzite (0001)	RMS (10:5)	
Co:ZnO 080312a(HV)	nomin. 10%	Al$_2$O$_3$ (sapphire)	wurtzite (0001)	RMS (10:5)	HV, 450°C 30 min.
Co:ZnO 080312a(O$_2$)	nomin. 10%	Al$_2$O$_3$ (sapphire)	wurtzite (0001)	RMS (10:5)	O$_2$, 450°C 30 min.
Co:ZnO 080131	nomin. 10% PIXE: 9.5%	Al$_2$O$_3$ (sapphire)	wurtzite (0001)	RMS (10:1)	
Co:ZnO 080428	nomin. 10%	Al$_2$O$_3$ (sapphire)	wurtzite (0001)	RMS (10:1)	
Co:ZnO 080428a(HV)	nomin. 10%	Al$_2$O$_3$ (sapphire)	wurtzite (0001)	RMS (10:1)	HV, 450°C 30 min.
Co:ZnO S0003-2	5·10^{19} Co/cm^3	ZnO (0001)	wurtzite (0001)	Co$^+$ FIB implanted, ≈100 nm	
Co:ZnO S0003-3	1·10^{20} Co/cm^3	ZnO (0001)	wurtzite (0001)	Co+ FIB implanted, ≈100 nm	
Co:ZnO 071115	nomin. 10%	Al$_2$O$_3$ (sapphire)	wurtzite (0001)	RMS (10:0.6)	
Co:ZnO 071115a(HV)	nomin. 10%	Al$_2$O$_3$ (sapphire)	wurtzite (0001)	RMS (10:0.6)	HV, 450°C 30 min.
Co:ZnO 071115a(O$_2$)	nomin. 10%	Al$_2$O$_3$ (sapphire)	wurtzite (0001)	RMS (10:0.6)	O$_2$, 450°C 30 min.

APPENDIX E: SAMPLE OVERVIEW

type/ signature	Co content	substrate	structure	growth (Ar:O$_2$)	annealing
Co:ZnO 071026	nomin. 10%	Al$_2$O$_3$ (sapphire)	wurtzite (0001)	RMS (8:1)	
Co:ZnO 080922	nomin. 5%	Al$_2$O$_3$ (sapphire)	wurtzite (0001)	RMS (10:0.8)	
Co:ZnO 080917	nomin. 10%	Al$_2$O$_3$ (sapphire)	wurtzite (0001)	RMS (10:0.5)	
Co:ZnO RU-1%	nom. 1% AAS: 2.8%	none	wurtzite nanopowder (<12 nm>)	chem. vapor synthesis (CVS)	
Co:ZnO RU-5%	nom. 5% AAS 11%	none	wurtzite nanopowder (<18 nm>)	chem. vapor synthesis (CVS)	
Co:ZnO RU-10%	nom. 10%	none	wurtzite nanopowder (<18 nm>)	chem. vapor synthesis (CVS)	
Co:ZnO ZOC5164	nomin. 5%	r-Al$_2$O$_3$ (sapphire)	wurtzite r-plane(01-12)	pulsed laser deposition	
Co:ZnO SC042807B	nomin. 10% PIXE: 10.8%	Al$_2$O$_3$ (sapphire)	wurtzite (0001)	pulsed laser deposition, off axis	
Co:ZnO SC042807C	nomin. 10%	Al$_2$O$_3$ (sapphire)	wurtzite (0001)	pulsed laser deposition, off axis	

Appendix C

Preparation Chamber

The preparation chamber used for this work is equipped with in total 7 magnetrons. Four magnetrons for 2 inch targets are placed in one quarter-circle of the cylindrical chamber (figure C b)). Since only one of these sources can be used at a time Co:ZnO has to be sputtered from composite targets. A respective magnetron station is sketched in figure C a). The sputter gas Ar as well as the reactant O_2 can be injected through the chimney of the magnetron by an inlet close to the plasma. Purifier and mass flow controllers outside the chamber ensure the control of the gas composition (not yet installed in C d)).

For preparation the sample is fixed in a sample holder and mounted face down on a rotatable baseplate. Hereby each magnetron position can be accessed. Shadow masks can be placed on the baseplate for patterning of the samples.

During preparation the sample is heated from the backside across the sample holder [1]. The heaters themselves are mounted from the top by adjustable metal rods connected to a feedthrough for the thermo couple and current supply. Details of the heater system are given in appendix D.

All magnetrons are equipped with pneumatic shutters to assure precise control of the exposure times of the substrate.

In addition the chamber contains three more 1.5 inch magnetrons and an ECR plasma source. These magnetrons are mounted on a cluster flange suitable for simultaneous co-sputtering.

The system is pumped by an ion getter pump, a Ti sublimation pump and a turbo

---

[1]The sample holder can also be mounted on the baseplate up side up. This directly exposes the sample to the heat radiation, e.g. for annealing purposes.

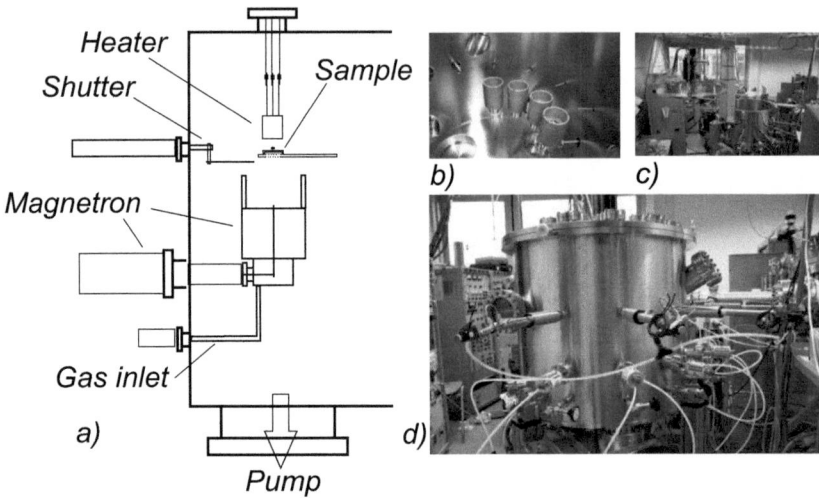

Figure C.1: a) Sketch of one magnetron sputter position in the preparation chamber. b) 4 magnetron stations c) open chamber with lifted top part d) magnetron preparation positions from the outside. Pictures were taken during installation of the system. Therefore not visible: Shutter plates in b) and heaters in c). Gas connections in d) are not installed yet.

pump with a scroll pump as prepump. The base pressure is in the low 10^{-9} mbar range. During preparation the first two pumps are isolated from the system by a gate valve and the working pressure ($\approx 10^{-4}$ mbar) is stabilized by mass flow controllers and a throttle valve to adjust the pumping speed of the turbo pump.

Appendix D

Heater system

Within this work a UHV-suitable sample heater system had to be developed. Mounting positions are given by the foreseen feedthroughs and preparation spots as described in C. Easy maintenance and reliability of the custom made mounting and double walled shielding for the heat radiation had to be considered.

The central component of the heating system consist of a commercial available Boralectric heater manufactured by the Advanced Ceramics Corporation retailed by TECTRA GmbH [1].

Specifications:

- vacuum compatibility (UHV compatible)
- very high temperatures up to 2.000 K
- chemically inert and dielectric surface
- very high heating rates ($>100°C/s$)

In particular being chemically very inert makes this heater suitable for reactive sputtering processes. Three thread hulls were welded on the flange of the feedthrough and onto the baseplate of the heater. Three 2 mm stainless steel rods with M2 threads could be connected to enable a 3 point stabilization. To assure a height adjustment each connection was built with two parallel rods guided through a connector. This connector was equipped with screws to fix the individual rod length. This construction guarantees the required height adjustment, within a small range a horizontal variability due to possible intended bending of the rods and a sufficient

---
[1] www.tectra.de

stiffness in case of movements (see also figure C a)). The latter is a major issue due to the chamber construction which had foreseen the removal of the whole top part of the chamber by a crane to have convenient access to all parts in case of a essentially target change (see figure C c)).

The baseplate fix both the heat shielding and the heater itself. The latter is connected via isolated tantalum rods that pass through the heat shielding. The double heat shielding itself consists of two components. The horizontal one is simply realized by a second stainless steel plate below the the baseplate, fixed isolated to the tantalum rods. The vertical component of the heat shielding consist of a two part double walled wrapping which can be very easily slipped over the base plate. This construction enables convenient removal of the vertical heat shielding and therefore access to all parts of the heater itself.

Four connections from the heater to the feedthrough had to be established. The current supply is realized by connectors at the tantalum rods. These rods therefore fulfill stabilization and function purposes, since the heater current is supplied through this rods. The second connection concenrns the temperature measurement direktly at the heater. It requires the stable positioning of a thermocouple in a 0.25 mm hole which was drilled horizontally into the heater to enable most precise temperature measurements (photo in D.1). The two wires of the thermocouple are guided through a 10 cm ceramic rod with two lead-throughs which extents from the heater to outside of the heat shielding, passing through a foreseen hole. To stabilize the horizontal position of the ceramic a little spring is connected to the heat shielding keeping the ceramic in place. Outside of the ceramic the stiffness of the K-type thermocouple is sufficient to guide the two wires separately that only the welding point inside the heater will contribute to the thermo voltage.

For calibration purposes a second thermocouple was installed to measure the temperature at the sample position. Due to the heat capacity of the sample holder,its connection to other metal parts, and radiation losses the sample position shows a reduced temperature. The temperature difference between heater and sample holder in the temperature range suitable for Co:ZnO preparation is typically 150 °C.

APPENDIX B: HEATER SYSTEM

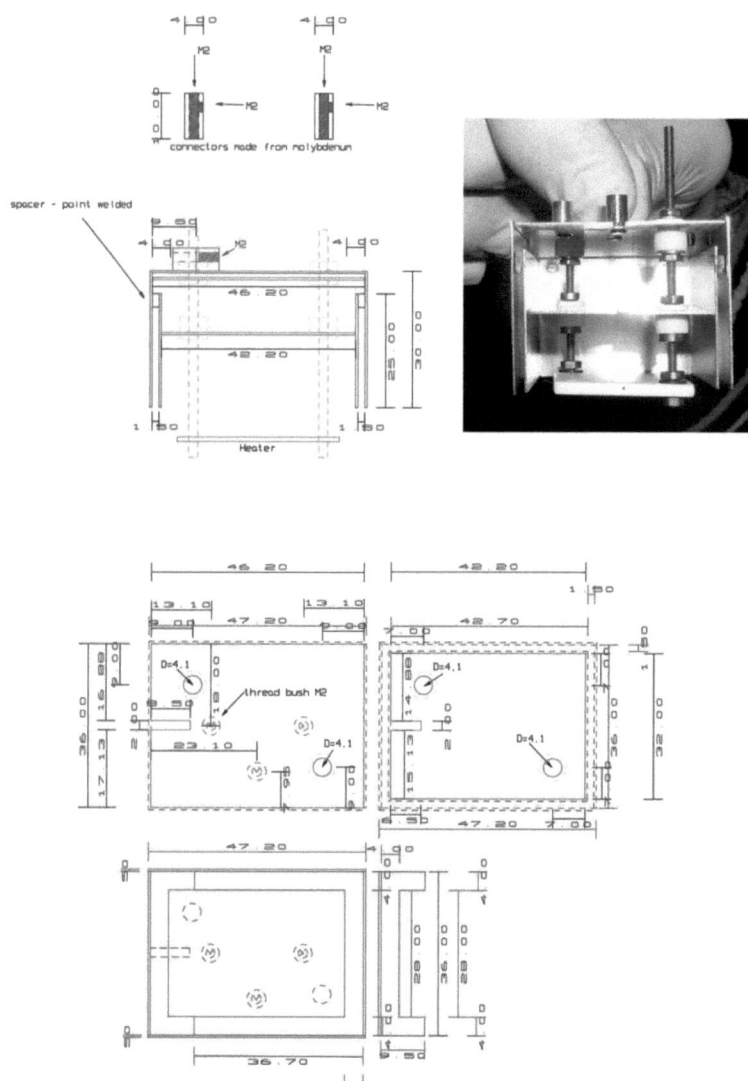

Figure D.1: Mounting and double shielding of heater. The photo shows the construction when the front part of the heat shielding is removed. In the center of the heater the foreseen hole for the thermocouple (not installed) is visible.

Appendix E

Bake out system

The preparation chamber described in appendix C is part of a connected UHV multi-chamber system which enables not only the preparation but also an in situ analysis of the samples. Part of this work was to develop a central bake-out control for three UHV chambers.

Requirements:

- power supply of up to 20 kW for three independent circuits
- pressure interlock for each circuit
- individual temperature control of up to 7 heater tape bunches within one circuit

15 commercially available digital temperature controllers [1] have been installed for individual control of the heater tape bunches. The temperature is measured by 15 J-type thermocouples fixed at respective positions of the UHV systems. Thus all relevant parts of the chambers can be set to specific temperatures during the bake-out procedure.

The heater tape bunches can be simply plugged in on the backside of the control box which offers 15 sockets corresponding to the temperature controllers. Additionally 6 uncontrolled sockets are foreseen to connect bake out systems with autonomous temperature control e.g. turbo- or ion pumps. All 21 sockets are individually fused. For each of the 3 bake-out circuits a pressure interlock is installed. Since every chamber is equipped with a separate pressure control, the respective relays of the

---
[1] ENDA ETC 1311

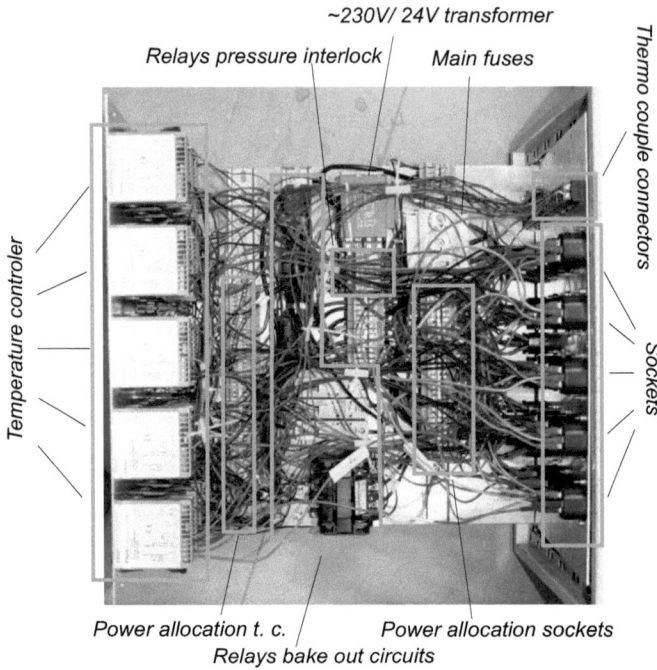

Figure E.1: Inner view of bake-out control

control units are used to interlock the bake-out circuits in case of extreme pressure increase. The control box needs a three phase current connection. The wiring of circuits explicitly accounts for a preferably equal weight on each phase.

APPENDIX C: BAKE OUT SYSTEM 183

Figure E.2: Front view of bake-out control

Figure E.3: back view of bake-out control

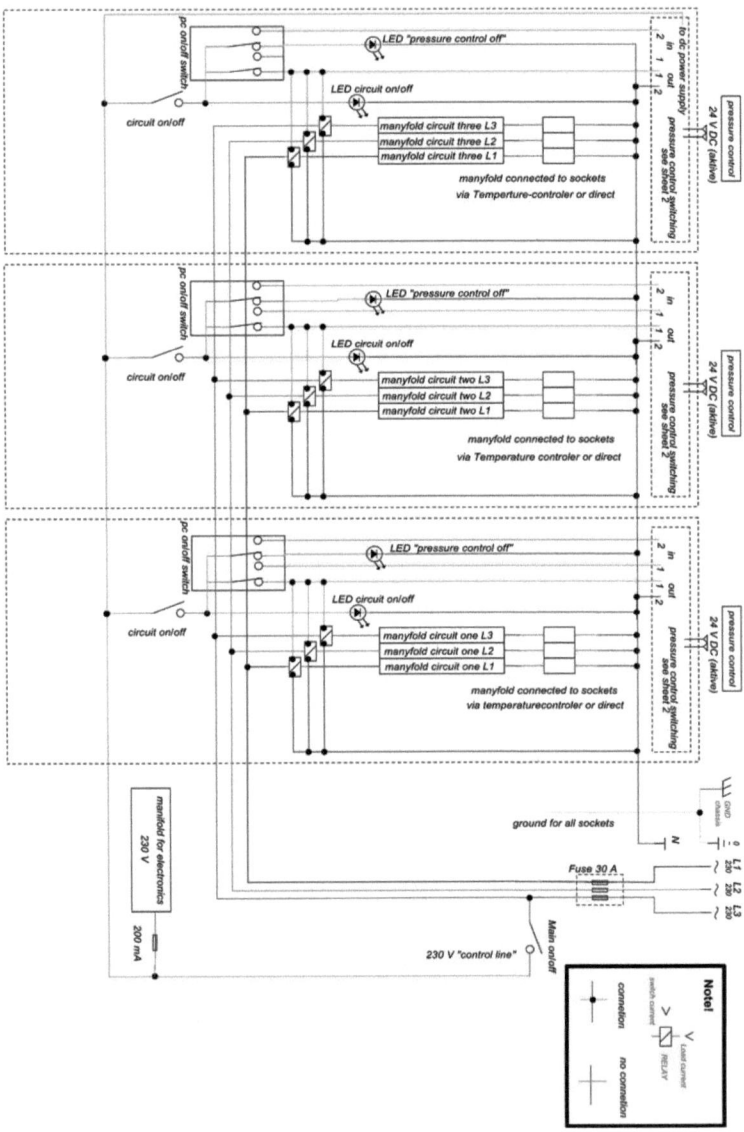

Figure E.4: Wiring diagram bake-out control

APPENDIX C: BAKE OUT SYSTEM

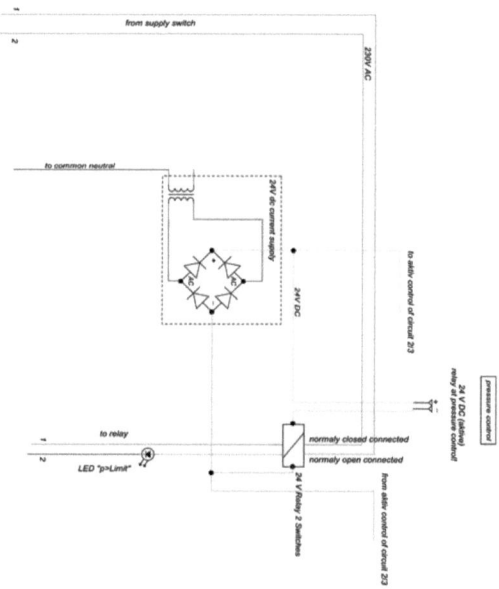

Figure E.5: Wiring diagram pressure interlock

Bibliography

[1] E. Grochowski and R. Hoyt, *Magnetics, IEEE Transactions on* **32**(3), 1850 (May 1996)

[2] J.-G. J. Zhu, *Materials Today* **6**(7-8), 22 (2003)

[3] M. N. Baibich, J. M. Broto, A. Fert, F. N. Van Dau, F. Petroff, P. Etienne, G. Creuzet, A. Friederich and J. Chazelas, *Phys. Rev. Lett.* **61**(21), 2472 (Nov 1988)

[4] G. Binasch, P. Grünberg, F. Saurenbach and W. Zinn, *Phys. Rev. B* **39**(7), 4828 (Mar 1989)

[5] S. Parkin, X. Jiang, K. R. Christian Kaiser, Alex Panchula and M. Samant, *Proceedings of the IEEE* **93**(5) (2003)

[6] T. Miyazaki and N. Tezuka, *Journal of Magnetism and Magnetic Materials* **151**(3), 403 (1995), Spin Polarized Electron Transport

[7] G. E. Moore, *Electronics* **38**(8) (1965)

[8] R. Schaller, *Spectrum, IEEE* **34**(6), 52 (Jun 1997)

[9] S. Datta and B. Das, *Applied Physics Letters* **56**(7), 665 (1990)

[10] S. A. Wolf, D. D. Awschalom, R. A. Buhrman, J. M. Daughton, S. von Molnar, M. L. Roukes, A. Y. Chtchelkanova and D. M. Treger, *Science* **294**(5546), 1488 (2001)

[11] A. Ney, C. Pampuch, R. Koch and K. H. Ploog, *Nature* **405**, 485 (2003)

[12] M. Johnson and R. H. Silsbee, *Phys. Rev. Lett.* **55**(17), 1790 (Oct 1985)

[13] M. Johnson, *Science* **260**(5106), 320 (1993)

[14] H. J. Zhu, M. Ramsteiner, H. Kostial, M. Wassermeier, H.-P. Schönherr and K. H. Ploog, *Phys. Rev. Lett.* **87**(1), 016601 (Jun 2001)

[15] A. T. Hanbicki, B. T. Jonker, G. Itskos, G. Kioseoglou and A. Petrou, *Applied Physics Letters* **80**(7), 1240 (2002)

[16] B. T. Matthias, R. M. Bozorth and J. H. Van Vleck, *Phys. Rev. Lett.* **7**(5), 160 (Sep 1961)

[17] A. Mauger and C. Godart, *Physics Reports* **141**(2-3), 51 (1986)

[18] A. H. MacDonald, P. Schiffer and N. T. Samarth, *Nature Materials* **4**, 195 (2005)

[19] H. Ohno, A. Shen, F. Matsukura, A. Oiwa, A. Endo, S. Katsumoto and Y. Iye, *Applied Physics Letters* **69**(3), 363 (1996)

[20] K. Y. Wang, R. P. Campion, K. W. Edmonds, M. Sawicki, T. Dietl, C. T. Foxon and B. L. Gallagher, *AIP Conf. Proc* **772**, 333 (2005)

[21] T. Dietl, H. Ohno, F. Matsukura, J. Cibert and D. Ferrand, *Science* **287**(5455), 1019 (2000)

[22] K. Sato and H. Katayama-Yoshida, *Japanese Journal of Applied Physics* **40**(Part 2, No. 4A), L334 (2001)

[23] N. Teraguchi, A. Suzuki, Y. Nanishi, Y.-K. Zhou, M. Hashimoto and H. Asahi, *Solid State Commun.* **122**, 651 (2002)

[24] A. Ney, K. Ollefs, S. Ye, T. Kammermeier, V. Ney, T. C. Kaspar, S. A. Chambers, F. Wilhelm and A. Rogalev, *Physical Review Letters* **100**(15), 157201 (2008)

[25] N. Akdogan, A. Nefedov, K. Westerholt, H. Zabel, H.-W. Becker, C. Somsen, R. Khaibullin and L. Tagirov, *Journal of Physics D: Applied Physics* **41**(16), 165001 (8pp) (2008)

[26] H.-J. Lee, S.-Y. Jeong, C. R. Cho and C. H. Park, *Applied Physics Letters* **81**(21), 4020 (2002)

[27] M. Venkatesan, C. B. Fitzgerald, J. G. Lunney and J. M. D. Coey, *Phys. Rev. Lett.* **93**(17), 177206 (Oct 2004)

[28] J. Stöhr and H. Siegmann, *Magnetism - From Fundamentals to Nanoscale Dynamics*, Springer (2008)

[29] B. Friedrich and D. Herschbach, *Physics Today* (12), 53 (2003)

[30] M. Farle and K. Baberschke, *Phys. Rev. Lett.* **58**(5), 511 (Feb 1987)

[31] R. Meckenstock, I. Barsukov, O. Posth, J. Lindner, A. Butko and D. Spoddig, *Applied Physics Letters* **91**(14), 142507 (2007)

[32] Y. Sasaki, X. Liu, J. K. Furdyna, M. Palczewska, J. Szczytko and A. Twardowski, Vol. 91, pp. 7484–7486. AIP (2002)

[33] F. Schwabl, *Statistische Mechanik*, Springer Berlin Heidelberg (2006)

[34] C. Rudowicz and H. W. F. Sung, *American Journal of Physics* **71**(10), 1080 (2003)

[35] C. Rudowicz, *arXiv:cond-mat/0304099v1* (2003)

[36] C. Kittel, *Einführung in die Festkörperphysik*, R. Oldenbourg Verlag, München, Wien (1999)

[37] W. Nolting, *Grundkurs Theoretische Physik 6: Statistische Physik*, Springer Verlag (2006)

[38] U. Özgür, Y. I. Alivov, C. Liu, A. Teke, M. A. Reshchikov, S. Doğan, V. Avrutin, S.-J. Cho and H. Morkoç, *Journal of Applied Physics* **98**(4), 041301 (2005)

[39] S. J. Pearton, C. R. Abernathy, M. E. Overberg, G. T. Thaler, D. P. Norton, N. Theodoropoulou, A. F. Hebard, Y. D. Park, F. Ren, J. Kim and L. A. Boatner, *Journal of Applied Physics* **93**(1), 1 (2003)

[40] H. J. Ko, Y. Chen, S. K. Hong and T. Yao, *Journal of Crystal Growth* **209**, 816 (2000)

[41] C. Liu, F. Yun and H. Morkoç, *Journal of Materials Science: Materials in Electronics* **16** (2005)

[42] Y. W. Zhang, X. M. Li, W. D. Yu, X. D. Gao, Y. F. Gu, C. Yang, J. L. Zhao, X. W. Sun, S. T. Tan, J. F. Kong and W. Z. Shen, *Journal of Physics D: Applied Physics* **41**(20), 205105 (6pp) (2008)

[43] A. Tsukazaki, A. Ohtomo, T. Onuma, M. Ohtani, T. Makino, M. Sumiya, K. Ohtani, S. F. Chichibu, S. Fuke, Y. Segawa, H. Ohno, H. Koinuma and M. Kawasaki, *Nature Mat.* **4** (2004)

[44] H. Qiu, B. Meyer, Y. Wang and C. Wöll, *Physical Review Letters* **101**(23), 236401 (2008)

[45] S. Strite and H. Morkoç, *Journal of Vacuum Science & Technology B: Microelectronics and Nanometer Structures* **10**(4), 1237 (1992)

[46] S. Hussain, *Investigation of Structural and Optical Properties of Nanocrystalline ZnO*, PhD Thesis, Linköpings Universtet, Sweeden (2008)

[47] R. C. Casella, *Phys. Rev.* **114**(6), 1514 (Jun 1959)

[48] R. Juza and H. Hahn, *Zeitschr. f. Anorg. Allgem. Chemie* **239**, 282 (1938)

[49] W. Paszkowicz, S. Podsiadlo and R. Minikayev, *Journal of Alloys and Compounds* **382**(1-2), 100 (2004), Proceedings of the European Materials Research Society Fall Meeting, Symposium B

[50] V. Darakchieva, B. Monemar and A. Usui, *Applied Physics Letters* **91**(3), 031911 (2007)

[51] C. P. Poole and H. A. Farach, *Theory of Magnetic Resonance*, Wiley-Intersience Publication (1987), 2nd edition

[52] A. Abragam and B. Bleaney, *Electron Paramagnetic Resonance of Transition Ions*, Calderon Press, Oxford (1970)

[53] H. Bethe, *Annalen der Physik* **395**, 133 (1929)

[54] R. O. Kuzian, A. M. Dare, P. Sati and R. Hayn, *Physical Review B* **74**, 155201 (2006)

[55] P. Koidl, *Phys. Rev. B* **15**(5), 2493 (Mar 1977)

[56] P. Sati, R. Hayn, R. Kuzian, S. Régnier, S. Schäfer, A. Stepanov, C. Morhain, C. Deparis, M. Laügt, M. Goiran and Z. Golacki, *Phys. Rev. Lett.* **96**(1), 017203 (2006)

[57] J. E. Wertz and J. R. Bolton, *Electron Spin Resonance*, Chapman and Hall Ltd (1972), reprint 1986

[58] J. A. Ibers and J. D. Swalen, *Phys. Rev.* **127**(6), 1914 (Sep 1962)

[59] C. Kittel, *Phys. Rev.* **71**(4), 270 (Feb 1947)

[60] C. Kittel, *Phys. Rev.* **73**(2), 155 (Jan 1948)

[61] M. Farle, *Rep. Prog. Phys.* **61**, 755 (1998)

[62] J. Lindner and M. Farle, *Magnetic Anisotropy of Heterostructures*, Vol. 227 of *Springer Tracts in Modern Physics*, Springer (2008)

[63] J. H. Van Vleck, *Phys. Rev.* **78**(3), 266 (May 1950)

[64] F. Bloch, *Phys. Rev.* **70**(7-8), 460 (Oct 1946)

[65] S. Dhar, O. Brandt, M. Ramsteiner, V. Sapega and K. Ploog, *Phys. Rev. Lett.* **94**(3) (JAN 28 2005)

[66] R. E. Behringer, *The Journal of Chemical Physics* **29**(3), 537 (1958)

[67] A. Bencini and D. Gatteschi, *EPR of Exchange Coupled Systems*, Springer-Verlag Berlin Heidelberg (1990)

[68] J. Hilgevoord, *American Journal of Physics* **64**(12), 1451 (1996)

[69] J. H. Van Vleck, *Phys. Rev.* **74**(9), 1168 (Nov 1948)

[70] M. Peric and B. L. Bales, *Journal of Magnetic Resonance* **169**(1), 27 (2004)

[71] J. J. Rehr and R. C. Albers, *Rev. Mod. Phys.* **72**(3), 621 (Jul 2000)

[72] R. Nakajima, J. Stöhr and Y. U. Idzerda, *Phys. Rev. B* **59**(9), 6421 (Mar 1999)

[73] V. Chakarian, Y. U. Idzerda and C. T. Chen, *Phys. Rev. B* **57**(9), 5312 (Mar 1998)

[74] W. L. O'Brien and B. P. Tonner, *Phys. Rev. B* **50**, 12672 (1994)

[75] P. Carra, B. T. Thole, M. Altarelli and X. Wang, *Phys. Rev. Lett.* **70**, 694 (1993)

[76] B. T. Thole, P. Carra, F. Sette and G. van der Laan, *Phys. Rev. Lett.* **68**(12), 1943 (Mar 1992)

[77] T. C. Kaspar, T. Droubay, S. M. Heald, P. Nachimuthu, C. M. Wang, V. Shutthanandan, C. A. Johnson, D. R. Gamelin and S. A. Chambers, *New Journal of Physics* **10**(5), 055010 (18pp) (2008)

[78] M. Winterer, V. V. Srdic, R. Djenadic, A. Kompch and T. E. Weirich, *Review of Scientific Instruments* **78**(12), 123903 (2007)

[79] J. Zenneck, *Optische Eigenschaften von verdünnten magnetischen Halbleitern auf GaN-Basis*, PhD Thesis (2007), Universität Göttingen

[80] L. Pérez, G. S. Lau, S. Dhar, O. Brandt and K. H. Ploog, *Physical Review B (Condensed Matter and Materials Physics)* **74**(19), 195207 (2006)

[81] F. Ziegler and J. P. Biersack, http://www.srim.org (2008), as at 15th January

[82] K. Seeger, *Semiconductor Physics*, Springer (1973)

[83] E. A. Ohm, *Ire transactions on microwave theory and techniques* pp. 210–217 (Feb 1956)

[84] J. D. Jackson, *Classical Electrodynamics*, John Wiley & Sons, New York (1975), 2nd edition

[85] C. P. Poole, *Electron Spin Resonance - A Comprehensive Treatise on Experimental Techniques*, John Wiley & Sons (1983), 2nd edition

[86] J. Wheatley and D. Halliday, *Phys. Rev.* **75**(9), 1412 (May 1949)

[87] A. Ney, T. Kammermeier, V. Ney, K. Ollefs and S. Ye, *JMMM* **320**, 3341 (2008)

[88] A. Rogalev and W. Wilhelm, http://www.esrf.eu/usersandscience/experiments/xasms/id12/ (2008), as at 10th January

[89] A. Thompson, D. Vaughan, J. Kirz, D. Attwood, E. Gullikson, M. Howells, K.-J. Kim, J. Kortright, I. Lindau and P. Pianetta, *X-ray data booklet*, Center for X-ray Optics and Advanced Light Source - Lawrence Berkeley National Laboratory (2001)

[90] J. Goulon, A. Rogalev, C. Gauthier, C. Goulon-Ginet, S. Paste, R. Signorato, C. Neumann, L. Varga and C. Malgrange, *Journal of Synchrotron Radiation* **5**(3), 232 (May 1998)

[91] Y. Joly, *Phys. Rev. B* **63**(12), 125120 (Mar 2001)

[92] A. Miheli, *XANES spectroscopy*, Univerza v Novi Gorici (2002)

[93] K. Ollefs, *Diploma thesis*, University of Duisburg-Essen (2008)

[94] E. H. Kisi and M. M. Elcombe, *Acta Crystallographica Section C* **45**(12), 1867 (Dec 1989)

[95] M. O. Krause and J. H. Oliver, *Journal of Physical and Chemical Reference Data* **8**(2), 329 (1979)

[96] S. Dhar, L. Pérez, O. Brandt, A. Trampert, K. H. Ploog, J. Keller and B. Beschoten, *Physical Review B (Condensed Matter and Materials Physics)* **72**(24), 245203 (2005)

[97] CrysTec company, Datasheet, provided after request

[98] S. Greulich-Weber, *Physica Status Solidi Applied Research* **162**, 95 (Juli 1997)

[99] E. N. Kalabukhova, N. N. Kabdin and S. Lukin, *Soviet Phys. - Solid State* **29**(8) (1987)

[100] C. F. Young, K. Xie, E. H. Poindexter, G. J. Gerardi and D. J. Keeble, *Applied Physics Letters* **70**(14), 1858 (1997)

[101] J. Isoya, T. Umeda, N. Mizuochi, N. T. Son, E. Janzén and T. Ohshima, *phys. stat. sol. (a)* **245**(7), 1298 (2008)

[102] T. Kammermeier, S. Dhar, V. Ney, E. Manuel, A. Ney, K. H. Ploog, F.-Y. Lo, A. Melnikov and A. D. Wieck, *phys. stat. sol. (a)* **205**(8), 1872 (2008)

[103] P. B. Dorain, *Phys. Rev.* **112**(4), 1058 (Nov 1958)

[104] W. M. Walsh and L. W. Rupp, *Phys. Rev.* **126**(3), 952 (May 1962)

[105] H.-J. Schulz and M. Thiede, *Phys. Rev. B* **35**(1), 18 (Jan 1987)

[106] D. D. Awschalom, J. Warnock and S. von Molnár, *Phys. Rev. Lett.* **58**(8), 812 (Feb 1987)

[107] N. Jedrecy, H. J. von Bardeleben, Y. Zheng and J.-L. Cantin, *Phys. Rev. B* **69**(4), 041308 (Jan 2004)

[108] W. C. Holton, J. Schneider and T. L. Estle, *Phys. Rev.* **133**(6A), A1638 (Mar 1964)

[109] G. W., Ludwig and H. H. Woodbury, *Solid State Physics* **13**, 223 (1962)

[110] T. Estle and M. D. Wit, *Bull. Am. Phys. Soc.* **6**, 445 (1961)

[111] P. Sati, A. Stepanov and V. Pashchenko, *Low Temperature Physics* **33**(11), 927 (2007)

[112] A. Einstein, *Verhandlungen der Deutschen Physikalischen Gesellschaft* **13/14**, 318 (1916)

[113] M. Farle and K. Baberschke, *Phys. Rev. Lett.* **58**(5), 511 (Feb 1987)

[114] A. R. S. Biasi and D. C. S. Rodrigues, *Journal of the American Ceramic Society* **68**(7), 409 (1985)

[115] K. N. Shrivastava, *Physics Reports* **20**(3), 137 (1975)

[116] K. J. Standley and R. A. Vaughan, *Phys. Rev.* **139**(4A), A1275 (Aug 1965)

[117] J. R. Lyons, *Telecommunications and Data Acquisition Progress Report* **98**, 63 (April 1989)

[118] M. O. Scheika-Kresimon, *Optisch detektierte paramagnetische Resonanz Spektroskopie am Rubin*, PhD Thesis, Universität Dortmund (2001)

[119] L. V. Nikolskaya, V. M. Terekhova and M. I. Samoilovich, *Physics and Chemistry of Minerals* **3**, 213 (1978)

[120] E. G. Sharoyan, O. S. Torosyan, E. A. Markosyan and V. T. Gabrielyan, *physica status solidi (b)* **65**(2), 773 (1974)

[121] D. I. Bletskan, V. Y. Bratus, A. R. Luk´yanchuk, V. T. Maslyuk and O. A. Parlag, *Technical Physics Letters* **34**(7), 612 (July 2008)

[122] M. Roever, D.-D. Mai, A. Bedoya-Pinto, J. Malindretos and A. Rizzi, *physica status solidi (c)* **5**(6), 2352 (2008)

[123] N. Grandjean, J. Massies, P. Vennéguès, M. Leroux, F. Demangeot, M. Renucci and J. Frandon, *Journal of Applied Physics* **83**(3), 1379 (1998)

[124] F.-Y. Lo, A. Melnikov, D. Reuter, A. D. Wieck, V. Ney, T. Kammermeier, A. Ney, J. Schörmann, S. Potthast, D. J. As and K. Lischka, *Applied Physics Letters* **90**(26), 262505 (2007)

[125] S. Dhar, T. Kammermeier, A. Ney, L. Pérez, K. H. Ploog, A. Melnikov and A. D. Wieck, *Applied Physics Letters* **89**(6), 062503 (2006)

[126] C. Liu, B. Mensching, M. Zeitler, K. Volz and B. Rauschenbach, *Phys. Rev. B* **57**(4), 2530 (Jan 1998)

[127] B. M. Ludbrook, The magnetic properties of selected rare earth nitrides grown by pulsed laser deposition (2009), Master's thesis, Victoria University of Wellington

[128] A. Hausmann, *Solid State Communications* **7**(8), 579 (1969)

[129] M. A. Khaderbad, S. Dhar, L. Pérez, K. H. Ploog, A. Melnikov and A. D. Wieck, *Applied Physics Letters* **91**(7), 072514 (2007)

[130] W. E. Carlos, J. A. Freitas, M. A. Khan, D. T. Olson and J. N. Kuznia, *Phys. Rev. B* **48**(24), 17878 (Dec 1993)

[131] R. Zeisel, M. W. Bayerl, S. T. B. Goennenwein, R. Dimitrov, O. Ambacher, M. S. Brandt and M. Stutzmann, *Phys. Rev. B* **61**(24), R16283 (Jun 2000)

[132] Z. Lipińska et al., *ACTA PHYSICA POLONICA A* **110**(2), 243 (2006)

[133] M. Fanciulli, T. Lei and T. D. Moustakas, *Phys. Rev. B* **48**(20), 15144 (Nov 1993)

[134] A. Ney, T. Kammermeier, E. Manuel, V. Ney, S. Dhar, K. H. Ploog, F. Wilhelm and A. Rogalev, *Applied Physics Letters* **90**(25), 252515 (2007)

[135] A. Rogalev, J. Goulon, C. Goulon-Ginet and C. Malgrange ()

[136] A. A. Freeman, K. W. Edmonds, G. van der Laan, R. P. Campion, A. W. Rushforth, N. R. S. Farley, T. K. Johal, C. T. Foxon, B. L. Gallagher, A. Rogalev and F. Wilhelm, *Physical Review B (Condensed Matter and Materials Physics)* **77**(7), 073304 (2008)

[137] A. Ney, T. Kammermeier, V. Ney, S. Ye, K. Ollefs, E. Manuel, S. Dhar, K. H. Ploog, E. Arenholz, F. Wilhelm and A. Rogalev, *Physical Review B* **77**(23), 233308 (2008)

[138] A. L. Patterson, *Phys. Rev.* **56**(10), 978 (Nov 1939)

[139] A. Rizzi, *priv. correspondence with A. Ney* (2008)

[140] C. Mitra and W. R. L. Lambrecht, *Physical Review B (Condensed Matter and Materials Physics)* **78**(13), 134421 (2008)

[141] F. Heigl, *Magnetooptics in Lanthanides*, PhD Thesis (2003), Freie Universität Berlin

[142] C.-G. Duan, R. F. Sabiryanov, W. N. Mei, P. A. Dowben, S. S. Jaswal and E. Y. Tsymbal, *Applied Physics Letters* **88**(18), 182505 (2006)

[143] T. C. Kaspar, T. Droubay, S. M. Heald, M. H. Engelhard, P. Nachimuthu and S. A. Chambers, *Physical Review B (Condensed Matter and Materials Physics)* **77**(20), 201303 (2008)

[144] M. Winterer, *Nanocrystalline Ceramics: Synthesis and Structure*, Springer, Heidelberg (2002)

[145] K. R. Kittilstved, D. A. Schwartz, A. C. Tuan, S. M. Heald, S. A. Chambers and D. R. Gamelin, *Physical Review Letters* **97**(3), 037203 (2006)

[146] T. Kammermeier, V. Ney, S. Ye, K. Ollefs, T. C. Kaspar, S. A. Chambers, F.Wilhelm, A. Rogalev and A. Ney, *JMMM* **321**(-), 699 (2009)

[147] A. Barla, G. Schmerber, E. Beaurepaire, A. Dinia, H. Bieber, S. Colis, F. Scheurer, J.-P. Kappler, P. Imperia, F. Nolting, F. Wilhelm, A. Rogalev, D. Müller and J. J. Grob, *Physical Review B (Condensed Matter and Materials Physics)* **76**(12), 125201 (2007)

[148] C. Kittel and E. Abrahams, *Phys. Rev.* **90**(2), 238 (Apr 1953)

[149] V. Ney, S. Ye, T. Kammermeier, A. Ney, H. Zhou, J. Fallert, H. Kalt, F.-Y. Lo, A. Melnikov and A. D. Wieck, *Journal of Applied Physics* **104**(8), 083904 (2008)

[150] U. Wiedwald, M. Spasova, E. L. Salabas, M. Ulmeanu, M. Farle, Z. Frait, A. F. Rodriguez, D. Arvanitis, N. S. Sobal, M. Hilgendorff and M. Giersig, *Phys. Rev. B* **68**(6), 064424 (Aug 2003)

[151] W. H. Meiklejohn and C. P. Bean, *Phys. Rev.* **102**(5), 1413 (Jun 1956)

[152] U. Wiedwald, M. Spasova, M. Farle, M. Hilgendorff and M. Giersig, Vol. 19, pp. 1773–1776. AVS (2001)

[153] R. Salzer, D. Spemann, P. Esquinazi, R. Höhne, A. Setzer, K. Schindler, H. Schmidt and T. Butz, *Journal of Magnetism and Magnetic Materials* **317**(1-2), 53 (2007)

[154] H. S. Hsu, J. C. A. Huang, Y. H. Huang, Y. F. Liao, M. Z. Lin, C. H. Lee, J. F. Lee, S. F. Chen, L. Y. Lai and C. P. Liu, *Applied Physics Letters* **88**(24), 242507 (2006)

[155] K. Rode, A. Anane, R. Mattana, J.-P. Contour, O. Durand and R. LeBourgeois, *Journal of Applied Physics* **93**(10), 7676 (2003)

[156] V. Ney, S. Ye, T. Kammermeier, K. Ollefs, A. Ney, T. Kaspar, S. Chambers, F. Wilhelm and A. Rogalev, *Journal of Magnetism and Magnetic Materials* pp. – (2009), In Press, Corrected Proof

[157] C. Antoniak, J. Lindner and M. Farle, *EPL (Europhysics Letters)* **70**(2), 250 (2005)

[158] H. J. von Bardeleben, N. Jedrecy and J. L. Cantin, *Applied Physics Letters* **93**(14), 142505 (2008)

[159] R. Galéra, J. Rueff, A. Rogalev, C. Giorgetti and E. Dartyge, *J. Synchrotron Rad.* (6), 676 (1999)

[160] D. P. Norton, M. E. Overberg, S. J. Pearton, K. Pruessner, J. D. Budai, L. A. Boatner, M. F. Chisholm, J. S. Lee, Z. G. Khim, Y. D. Park and R. G. Wilson, *Applied Physics Letters* **83**(26), 5488 (2003)

[161] J. M. D. Coey, M. Venkatesan and C. B. Fitzgerald, *Nature Materials* **4**(2) (2005)

[162] D. Chakraborti, G. R. Trichy, J. T. Prater and J. Narayan, *Journal of Physics D: Applied Physics* **40**(24), 7606 (2007)

[163] E. H. Williams, *Phys. Rev.* **28**(1), 167 (Jul 1926)

[164] J. Sakurai, W. J. L. Buyers, R. A. Cowley and G. Dolling, *Phys. Rev.* **167**(2), 510 (Mar 1968)

[165] M. Inada, *Japanese Journal of Applied Physics* **17**(1), 1 (1978)

[166] P. Dutta, M. S. Seehra, S. Thota and J. Kumar, *Journal of Physics: Condensed Matter* **20**(1), 015218 (8pp) (2008)

[167] A. H. J. Kim, I. C. Song, J. H. Sim, H. Kim, D. Kim, Y. E. Ihm and W. K. Choo, *physica status solidi (b)* **241**(7), 1553 (2004)

[168] K. Samanta, P. Bhattacharya, R. S. Katiyar, W. Iwamoto, P. G. Pagliuso and C. Rettori, *Physical Review B (Condensed Matter and Materials Physics)* **73**(24), 245213 (2006)

[169] W. Roth, *Journal of Physics and Chemistry of Solids* **25**, 1 (Januar 1964)

[170] M. Sikora, C. Kapusta, K. Knížek, Z. Jirák, C. Autret, M. Borowiec, C. J. Oates, V. Procházka, D. Rybicki and D. Zajac, *Physical Review B (Condensed Matter and Materials Physics)* **73**(9), 094426 (2006)

[171] Y. Zhang, Y. Chen, T. Wang, J. Zhou and Y. Zhao, *Microporous and Mesoporous Materials* **114**(1-3), 257 (2008)

[172] L. Hu, Q. Peng and Y. Li, *Journal of the American Chemical Society* **130**(48), 16136 (2008)

[173] T. Koide, H. Miyauchi, J. Okamoto, T. Shidara, A. Fujimori, H. Fukutani, K. Amemiya, H. Takeshita, S. Yuasa, T. Katayama and Y. Suzuki, *Phys. Rev. Lett.* **87**(25), 257201 (Nov 2001)

[174] H. L. Meyerheim, C. Tusche, A. Ernst, S. Ostanin, I. V. Maznichenko, K. Mohseni, N. Jedrecy, J. Zegenhagen, J. Roy, I. Mertig and J. Kirschner, *Physical Review Letters* **102**(15), 156102 (2009)

[175] A. S. Risbud, L. P. Snedeker, M. M. Elcombe, A. K. Cheetham and R. Seshadri, *Chem. Mater.* **17**(4), 834 (2005)

[176] J. Alaria, N. Cheval, K. Rode, M. Venkatesan and J. M. D. Coey, *Journal of Physics D: Applied Physics* **41**(13), 135004 (6pp) (2008)

[177] G. R. Hanson, K. E. Gates, C. J. Noble, M. Griffin, A. Mitchell and S. Benson, *Journal of Inorganic Biochemistry* **98**(5), 903 (2004), Contributions from the 11th International Conference on Biological Inorganic Chemistry

[178] A. Sukhov, K. Usadel and U. Nowak, *Journal of Magnetism and Magnetic Materials* **320**(1-2), 31 (2008)

[179] M. Snure, D. Kumar and A. Tiwari, *JOM Journal of the Minerals, Metals and Materials Society* **61**(6) (2009)

[180] Y. J. Li, T. C. Kaspar, T. C. Droubay, A. G. Joly, P. Nachimuthu, Z. Zhu, V. Shutthanandan and S. A. Chambers, *Journal of Applied Physics* **104**(5), 053711 (2008)

[181] A. J. Behan, A. Mokhtari, H. J. Blythe, D. Score, X.-H. Xu, J. R. Neal, A. M. Fox and G. A. Gehring, *Physical Review Letters* **100**(4), 047206 (2008)

[182] S. Ye, V. Ney, T. Kammermeier, K. Ollefs, S. Zhou, H. Schmidt, F. Wilhelm, A. Rogalev and A. Ney (), submitted to PRB

[183] E. P. Wigner, *Group Theory and its application to the quantum mechanics of atomic spectra*, Acacemic Press INC. (London) (1959)

[184] J. S. Griffith, *The theory of transition metal ions*, University press, Cambridge (1961)

[185] R. M. Macfarlane, *Phys. Rev. B* **1**(3), 989 (Feb 1970)

[186] F. S. Ham, G. W. Ludwig, G. D. Watkins and H. H. Woodbury, *Phys. Rev. Lett.* **5**(10), 468 (Nov 1960)

[187] P. Sati, *Propriétés magnétiques et de résonance du $Zn_{1-x}Co_xO$: un matériau candidat pour l'électronique de spin*, PhD Thesis (2007), Universitité Paul Cézanne (Aix-Marseille III)

[188] J. M. Baranowski, J. W. Allen and G. L. Pearson, *Phys. Rev.* **160**(3), 627 (Aug 1967)

[189] G. Racah, *Phys. Rev.* **62**(9-10), 438 (Nov 1942)

[190] S. Isber, M. Averous, Y. Shapira, V. Bindilatti, A. N. Anisimov, N. F. Oliveira, V. M. Orera and M. Demianiuk, *Phys. Rev. B* **51**(21), 15211 (Jun 1995)

List of publications

as at 1. October 2009

1. Christoph Knies, Matthias T. Elm, Peter J. Klar, Jan Stehr, Detlev M. Hofmann, Nikolai Romanov, Tom Kammermeier and Andreas Ney
   Nonferromagnetic nanocrystalline ZnO:Co thin films doped with Zn interstitials
   J. Appl. Phys. **105**, 073918 (2009)

2. A. Ney, T. Kammermeier, K. Ollefs, V. Ney, S. Ye, S. Dhar, K. H. Ploog, M. Röver, J. Malindretos, A. Rizzi, F. Wilhelm, A. Rogalev
   Gd-doped GaN studied with element specifity: very small polarization of Ga, paramagnetism of Gd and the formation of magnetic clusters
   J. Magn. Magn. Mat. (2009), doi: 10.1016/j.jmmm.2009.06.033

3. V. Ney, S. Ye, T. Kammermeier, K. Ollefs, A. Ney, T.C. Kaspar, S.A. Chambers, F. Wilhelm and A. Rogalev
   Tuning the magnetic properties of $Zn_{1-x}Co_xO$ films
   J. Magn. Magn. Mat. (2009), DOI 10.1016/j.jmmm.2009.04.024

4. T. Kammermeier, V. Ney, S. Ye, K. Ollefs, T. C. Kaspar, S. A. Chambers, F. Wilhelm, A. Rogalev, and A. Ney
   Element specific measurements of the structural properties and magnetism of $Co_xZn_{1-x}O$
   J. Magn. Magn. Mater. **321**, 699 (2009)

5. V. Ney, S. Ye, T. Kammermeier, A. Ney, H. Zhou, J. Fallert, H. Kalt, F.-Y. Lo, A. Melnikov, and A. D. Wieck
   Structural, Magnetic and Optical Properties of Co- and Gd-implanted ZnO(0001) substrates
   J. Appl. Phys. **104**, 083904 (2008)

6. A. Ney, T. Kammermeier, V. Ney, K. Ollefs, and S. Ye
   Limitations of measuring small magnetic signals of samples deposited on a diamagnetic substrate
   J. Magn. Magn. Mater. **320**, 3341 (2008)

7. T. Kammermeier, S. Dhar, V. Ney, E. Manuel, A. Ney, K. H. Ploog, F.-Y. Lo, A. Melnikov, A. D. Wieck
   Paramagnetic and ferromagnetic resonance studies on dilute magnetic semiconductors based on GaN
   phys. stat. sol. (a) **205**(8), 1872 (2008)

8. A. Ney, R. Rajaram, T. Kammermeier, V. Ney, S. Dhar, K.H. Ploog and S.S.P. Parkin
   Metastable magnetism and memory effects in dilute magnetic semiconductors
   J. Phys.: Condens. Matter **20**, 285222 (2008)

9. A. Ney, T. Kammermeier, V. Ney, S. Ye, K. Ollefs, E. Manuel, S. Dhar, K. H. Ploog, E. Arenholz, F. Wilhelm, and A. Rogalev
   Element specific magnetic properties of Gd:GaN: very small polarization of Ga and paramagnetism of Gd
   Phys. Rev. B **77**, 233308 (2008)

10. A. Ney, K. Ollefs, S. Ye, T. Kammermeier, V. Ney, T. C. Kaspar, S. A. Chambers, F. Wilhelm, and A. Rogalev
    Absence of intrinsic ferromagnetic interactions of isolated and paired Co dopant atoms in $Zn_{1-x}Co_xO$ with high structural perfection
    Phys. Rev. Lett. **100**, 157201 (2008)

11. A. Ney, R. Rajaram, S.S.P. Parkin, T. Kammermeier, S. Dhar
    Experimental investigation of the metastable magnetic properties of Cr-doped InN
    Phys. Rev. B **76**, 035205 (2007)

LIST OF PUBLICATIONS

12. F.-Y. Lo, A. Melnikov, D. Reuter, A. D. Wieck, V. Ney, T. Kammermeier, A. Ney, J. Schörmann, S. Potthast, D. J. As, and K. Lischka
    Magnetic and structural properties of Gd-implanted zinc-blende GaN
    Appl. Phys. Lett. **90**, 262505 (2007)

13. A. Ney, T. Kammermeier, E. Manuel, V. Ney, S. Dhar, K.H. Ploog, F. Wilhelm, and A. Rogalev
    Element specific investigations of the structural and magnetic properties of Gd:GaN
    Appl. Phys. Lett. **90**, 252515 (2007)

14. A. Ney, R. Rajaram, S. S. P. Parkin, T. Kammermeier, S. Dhar
    Magnetic properties of epitaxial CrN films
    Appl. Phys. Lett. **89**, 112504 (2006)

15. S. Dhar, T. Kammermeier, A. Ney, L. Perez, K. H. Ploog, A. Melnikov, and A. D. Wieck
    Ferromagnetism and colossal magnetic moment in Gd-focused ion-beam-implanted GaN
    Appl. Phys. Lett. **89**, 062503 (2006)

List of acronyms

AAS:	Atomic Absorption Spectroscopy
CVS:	Chemical Vapor Synthesis
DMS:	Dilute Magnetic Semiconductor
EDXD:	Energy Dispersive X-ray Diffraction
EPR:	Electron Paramagnetic Resonance
ESR:	Electron Spin Resonance
ESRF:	European Synchrotron Radiation Facility
EXAFS:	Extended X-ray Absorption Fine Structure
FIB:	Focused-Ion-Beam
FMR:	Ferromagnetic Resonance
FWHM:	Full Width at Half Maximum
GMR:	Giant Magnetoresistance
MBE:	Molecular Beam Epitaxy
MOCVD:	Metal-Organic Chemical Vapor Deposition
PIXE:	Proton Induced X-ray Emission
PLD:	Pulsed Laser Deposition
RKKY:	Ruderman-Kittel-Kasuya-Yosida
RMS:	Reactive Magnetron Sputtering
SQUID:	Superconducting Quantum Interference Device
TMR:	Tunneling Magnetoresistance
XANES:	X-ray Absorption Near Edge Spectroscopy
XAS:	X-ray Absorption Spectroscopy
XLD:	X-ray Linear Dichroism
XMCD:	X-ray Magnetic Circular Dichroism
XRD:	X-ray Diffraction

Acknowledgement

This work would have been impossible without continuous support from many people. I would like to express my gratitude to all those who gave me the possibility to complete this thesis. In particular I want to thank:

- Prof. M. Farle for being the official PhD thesis supervisor and for his hospitality for the MAGLOMAT-Project.

- Dr. A. Ney, who gave me invaluable contributions by our discussions, his recommendations and his patient proof reading. Besides this I have to thank him for his friendship turning the last years of work into such a nice time.

- Dr. S. Ye not only for the fruitful scientific discussions, but also for the interesting insight into Chinese's thinking.

- Dr. V. Ney for her help with the XRD data and for insisting on the MAGLOMAT rules concerning accepted papers.

- K. Ollefs for asking so many questions – 'till showing me the limits of my knowledge...

- Prof. S. Dhar, I want to thank for his encouraging way to explain physics, which motivated me in particular for the work on Gd:GaN.

- Horst Zähres for his support on many technical aspects. If *Heinzelmännchen* exsist, he's one!

- Michael Vennemann for helping me keeping my Computer alive – which was sometimes not that easy!

- Dieter Schädel for his CAD lessons and turning my ideas into functioning constructions.

- Sabina Grubba and Helga Mundt for patiently correcting my filled forms and of course providing sweets during the last years...

- Florian Römer for providing not only very useful origin-tools, but also stress reducing magnetic gadgets...

- Dr. Caroline Antoniak and Dr. Anastassia Trunova for their help in the beginning of my work at the ESR/FMR machines.

- Dr. Fabrice Wilhelm and A. Rogalev for their support at the beamline ID12 at the ESRF.

- Prof. F.-Y. Lo, Prof. A. D. Wieck and collaborators for the ion implantation

- For making Gd:GaN samples available I have to thank the group of Prof K. H. Ploog and the group of Prof. A. Rizzi.

- Prof .S. Chambers and Dr. T. C. Kaspar I have to thank for providing PLD grown Co:ZnO.

- Dr. Milan Gacic and Prof. H. Adrian for providing a r-plane PLD grown Co:ZnO sample.

- Prof. D. Hofmann and colleagues for the nice stay in Gießen.

- and all other actual and former members of the AG Farle group: Seda Aksoy, Anja Bannholzer, Igor Barsukov, Anna Elsukova, Nina Friedenberger, Yu Fu, Prof. M. Acet, Christoph Hassel, Dr. Jochen Kästner, Wolfgang Kunze, Dr. Zi-An Li, Dr. Jürgen Lindner, Dr. Ralf Meckenstock, Vadim Migunov, Prof. G. Dumpich, Oliver Posth, Nathalie Reckers, Dr. Marina Spasova, Sven Stienen, Gökhan Ünlü, Christian Wirtz, Dr. Thomas Kebe, Dr. Olga Dmitrieva, Thomas Fridrich, Michaela Hartmann, Dr. Eyup Duman, Anastasia Konstantinova, Dr. Thorsten Krenke, Patryk Krzysteczko, Andrey Lysov, Esperanca Manuel, Dr. Olivier Margeat, Dr. Oliver Muth, Dr. Christian Raeder, Dr. Irina Rod, Zahra Shojaaee, Dr. Burkhard Stahlmecke, Dr. Daniela Sudfeld, Cihan Tomaz, Michael Tran, Dr. Khalil Zakeri

ACKNOWLEDGEMENT

...last but not least I have to thank Henrike and Jorin for showing me what really matters when coming home...

ACKNOWLEDGEMENT

Die VDM Verlagsservicegesellschaft sucht für wissenschaftliche Verlage abgeschlossene und herausragende

Dissertationen, Habilitationen, Diplomarbeiten, Master Theses, Magisterarbeiten usw.

für die kostenlose Publikation als Fachbuch.

Sie verfügen über eine Arbeit, die hohen inhaltlichen und formalen Ansprüchen genügt, und haben Interesse an einer honorarvergüteten Publikation?

Dann senden Sie bitte erste Informationen über sich und Ihre Arbeit per Email an *info@vdm-vsg.de*.

Sie erhalten kurzfristig unser Feedback!

VDM Verlagsservicegesellschaft mbH
Dudweiler Landstr. 99   Telefon   +49 681 3720 174
D - 66123 Saarbrücken   Fax   +49 681 3720 1749
www.vdm-vsg.de

Die VDM Verlagsservicegesellschaft mbH vertritt

Printed by Books on Demand GmbH, Norderstedt / Germany